Transport and Energy Processes 2020

Held at the 2020 AIChE Annual Meeting

Online
16 - 20 November 2020

ISBN: 978-1-7138-2302-5

Printed from e-media with permission by:

Curran Associates, Inc.
57 Morehouse Lane
Red Hook, NY 12571

Some format issues inherent in the e-media version may also appear in this print version.

Printed with permission by Curran Associates, Inc. (2021)

For permission requests, please contact AIChE
at the address below.

AIChE
120 Wall Street, FL 23
New York, NY 10005-4020

Phone: (800) 242-4363
Fax: (203) 775-5177

www.aiche.org

Additional copies of this publication are available from:

Curran Associates, Inc.
57 Morehouse Lane
Red Hook, NY 12571 USA
Phone: 845-758-0400
Fax: 845-758-2633
Email: curran@proceedings.com
Web: www.proceedings.com

TABLE OF CONTENTS

(391A) MODELING THE REACTIONS AND TRANSPORT IN GAS DIFFUSION ELECTRODE (GDE) FOR THE ELECTROCHEMICAL REDUCTION OF NITROGEN TO AMMONIA 1
Nishithan C. Kani, Meenesh R. Singh

(391C) METHOD OF MOMENTS TO CHARACTERIZE POROUS MEDIA USING NON-IDEAL TRACER 2
Deepshikha Singh, Jyoti Phirani

(391E) EVOLUTION AND INTERACTION CHARACTERISTICS OF LIQUID FLOW AND BUBBLES IN A JET BUBBLING COLUMN 3
Xiaoyun Guo, Zhengliang Huang, Jingyuan Sun, Yun Shuai, Yao Yang, Zuwei Liao, Jingdai Wang, Yongrong Yang

(398A) RATIONAL DESIGN OF SOLID-LIQUID INTERPHASES FOR HIGH-ENERGY SODIUM METAL BATTERIES 5
Shuya Wei

(398B) QUANTIFICATION OF LITHIUM PLATING AND SOURCES OF CAPACITY LOSS DURING FAST CHARGING OF GRAPHITE ELECTRODES 6
Eric McShane, Andrew Colclasure, David Brown, Zachary Konz, Kandler Smith, Bryan D. McCloskey

(398C) LITHIUM PLATING DETECTION DURING EXTREME FAST CHARGING IN LITHIUM-ION BATTERIES USING SIMULTANEOUS NEUTRON AND X-RAY IMAGING 7
Maha Yusuf, Jacob Lamanna, Partha Paul, Lambertus Hesselink, Michael Toney, Johanna Weker

(398D) SURFACE DEGRADATION OF LI-ION BATTERY CATHODES FROM FIRST PRINCIPLES 9
Xinhao Li, Alexander Urban

(398E) UNDERSTANDING ELECTROCHEMICAL REACTION MECHANISMS AND PROPERTIES OF RECHARGEABLE ALUMINUM-SULFUR AND ALUMINUM-SELENIUM BATTERIES 10
Robert Messinger, Rahul Jay, Ankur L. Jadhav

(398G) ELECTROCHEMICAL KINETICS OF DEGRADATION OF POROUS GRAPHITE ELECTRODES IN LITHIUM ION BATTERIES 11
Supratim Das, Martin Z. Bazant

(398H) STRATEGIES FOR PERFORMANCE IMPROVEMENT OF POLYOXOMETALATES IN NON-AQUEOUS REDOX FLOW BATTERIES 13
Yuan Cao, Jee-Jay J. Chen, Mark A. Barteau

(398I) BRIDGING MECHANISM OF V2+/V3+ REACTION ON GLASSY CARBON IN ACIDIC ELECTROLYTES FOR VANADIUM REDOX FLOW BATTERIES 14
Harsh Agarwal, Jacob Florian, Bryan R. Goldsmith, Nirala Singh

(398J) ELECTROCHEMICAL SYSTEMS OPTIMIZATION USING PHYSICS-BASED HOMOGENIZATION OF THE MICROSCOPIC CONTINUUM MODELS 17
Victor A. Beck, Seth E. Watts, Jonathan M. Wong, Daniel A. Tortorelli, Sarah Baker, Eric B. Duoss, Marcus Worsley

(461A) CFD SIMULATIONS OF TANGENTIALLY FIRED COAL BOILER UNDER VARIOUS LOADS 18
Yong Liu, Mehrdad Shahnam, Chris Guenther

(461B) MULTIOBJECTIVE CONTROL OF FIBER MORPHOLOGY IN PULPING PROCESS VIA MULTISCALE MODELING 19
Hyun-Kyu Choi, Sang Hwan Son, Joseph Kwon

(461C) WEAK FORMULATION OF SUPERCOOLED STEFAN PROBLEM FOR SOLIDIFICATION OF PURE SILICON IN HORIZONTAL RIBBON GROWTH 21
Eyan Noronha, B. Erik Ydstie

(461D) DESIGNING AND OPTIMIZING ENERGY POLICIES THROUGH A NOVEL ENERGY PRICE PREDICTIVE FRAMEWORK 23
Stefanos G. Baratsas, Alexander M. Niziolek, Onur Onel, Logan R. Matthews, Christodoulos A. Floudas, Detlef R. Hallermann, Sorin M. Sorescu, Efstratios N. Pistikopoulos

(461E) IDENTIFYING OPTIMAL KINETIC PATHWAYS OF A FUNCTIONALIZED SOLID SORBENT FOR CO2 CAPTURE THROUGH MATHEMATICAL PROGRAMMING 24
Ryan Hughes, Chinmay Aras, Goutham Kotamreddy, Debangsu Bhattacharyya

(461F) REACTION-PHASE EQUILIBRIUM ANALYSIS OF CH4, CO, CO2, H2O, H2, C IDEAL MIXTURES IN C, H, O ATOM SPACE USING GIBBS FREE ENERGY GLOBAL MINIMIZATION 26
Demetrios Chaconas, Patricia Pichardo, Ioannis Manousiouthakis, Vasilios Manousiouthakis

(461G) A NONSMOOTH APPROACH TO MULTICONTAMINANT WATER INTEGRATION 43
Caroline Nielsen, Paul I. Barton

(461H) A COMPREHENSIVE INVESTIGATION OF DENDRITE FORMATION IN LITHIUM ANODE BATTERIES: A COMPUTATIONAL APPROACH FOR HETEROGENOUS SEI 44
Niranjan Sitapure, Hyeonggeon Lee, Perla B. Balbuena, Sungwon Hwang, Joseph Kwon

(485D) PERFORMANCE AND DURABILITY OF ADVANED MN-BASED PRECIOUS-GROUP METAL-FREE CATALYST IN MEMBRANE ELECTRODE ASSEMBLIES 46
Fan Yang, Thomas Stracensky, Sichen Zhong, Mengjie Chen, Lin Guo, Yun Wang, Gang Wu, Hui Xu

(485E) ACHIEVING HIGH PEFC PERFORMANCE AT RATED POWER THROUGH MODIFICATIONS TO ELECTRODE MICROSTRUCTURE 47
Timothy Van Cleve, Firat Cetinbas, Guanxiong Wang, Mason Mooney, Nancy N. Kariuki, Deborah J. Myers, K. C. Neyerlin

(485F) DURABLE AND HIGH-POWER PGM-FREE CATHODES FOR PEM FUEL CELLS 48
Gang Wu

(518A) MICROELECTRODE STUDY OF OXYGEN MASS TRANSPORT AND ITS INFLUENCE ON THE PERFORMANCE OF HIGH TEMPERATURE PROTON EXCHANGE MEMBRANE FUEL CELLS 49
Shuomeng Zhang, Jujia Zhang, Shanfu Lu, Qinggang He

(518B) IMPACT OF IONOMER CONTENT ON FUEL-CELL CATALYST LAYER STRUCTURE 50
Aaron He, Nick Dagan, Anamika Chowdhury, Sarah A. Berlinger, Clayton J. Radke, Adam Z. Weber

(518D) MEASURING INTERNAL RESISTANCE OF A MICROBIAL FUEL CELL 52
Luay Ahmed, William Sweeny, Patrick Metz, Benjamin Daniels, Alim Dewan

(518F) MODELING THE CAPACITY LOSS OF AN AGM LEAD-ACID BATTERY 53
Jaewoo Lee, Dongcheul Lee, Byungmook Kim, Chee Burm Shin

(518H) INFLUENCE OF ADDITIVES ON THE PERFORMANCE AND FAILURE MECHANISMS OF THE RECHARGEABLE ALKALINE ZINC ELECTRODE 54
Michael D'Ambrose, Damon E. Turney, Gautam G. Yadav, Michael Nyce, Robert Messinger, Sanjoy Banerjee

(518J) ELECTROCHEMICAL TECHNIQUES TO DETECT AND QUANTIFY LI PLATING AFTER FAST CHARGE OF LI-ION BATTERIES 55
Zachary Konz

(518K) REDUCING ZINC REDISTRIBUTION AND EXTENDING CYCLE LIFE WITH ELECTROCHEMICALLY SYNTHESIZED ZINC OXIDE ANODES IN RECHARGEABLE ALKALINE BATTERIES 56
Snehal Kolhekar, Michael Nyce, Sanjoy Banerjee

(518L) PERSISTENT, SINGLE-POLARITY ENERGY HARVESTING FROM AMBIENT THERMAL FLUCTUATIONS USING A THERMAL RESONANCE DEVICE WITH THERMAL DIODES 58
Ge Zhang, Anton L. Cottrill, Volodymyr Koman, Albert Tianxiang Liu, Michael Strano

(518M) PROBING THE LINK BETWEEN AQUEOUS-PHASE CE3+/CE4+ STRUCTURE AND REDOX KINETICS FOR ENERGY STORAGE 59
Cailin Buchanan, Eunbyeol Ko, Bryan R. Goldsmith, Nirala Singh

(518P) DETAILED MODELLING AND SIMULATION OF THERMOPHYSICAL PROPERTY BASED HYDROGEN LIQUEFACTION PROCESS WITH EXERGY AND ECONOMIC ANALYSIS 60
Dongjun Lee, Sungwon Hwang

(555A) DOWNSTREAM PROCESS WITH COUNTERCURRENT CHROMATOGRAPHY TO RECOVER BIOPRODUCTS FROM SUSTAINABLE FEEDSTOCKS 61
Hoon Choi, Patrick Saboe, Brenna Black, Dong Xueming, Stefan Haugen, Eric M. Karp

(555B) SEPARATION OF LIGNOCELLULOSIC SUGARS FROM PYROLYTIC BIO-OIL USING SIMULATED MOVING BED 62
Arpa Ghosh, John P. Stanford, Ryan G. Smith, Robert C. Brown

(555C) BLACK LIQUOR GASIFICATION CHEMICAL RECOVERY AND COMBINED CYCLE POWER AT PULP AND PAPER MILLS BIOREFINERY 63
Tapas K. Das

(555D) FROM BIOMASS TO POWER USING INTEGRATED COMPACT PYROLYSIS, COMBUSTOR, AND STIRLING ENGINE 64
Sampath Gunukula, Dat T. Tran

(555E) TECHNO-ECONOMIC ANALYSIS AND LIFE-CYCLE ASSESSMENT OF EMERGING TECHNOLOGIES FOR BIOPROCESSING SEPARATIONS 65
Lauren Valentino, Jennifer B. Dunn, Eric Tan, Charles J. Freeman, William L. Kubic Jr., Alex Rosenthal

(568A) PROMOTED EFFECT OF MALEIC ACID AS AN ADDITIVE ON THERMAL DEHYDROGENATION OF AMMONIA BORANE 66
Geo-Jong Kim, Savannah G. Hunt, Hyun-Tae Hwang

(568B) THE PRE-ACIDIFICATION GAS IMPACT ON UPGRADING THE BIOGAS PRODUCED IN EXPANDED GRANULAR SLUDGE BED REACTOR 67
Haider Al-Rubaye, Joseph D. Smith, Jia Yu, Mahyar Ghorbanian, Manohar Manchenahalli, Anand Alembath

(568C) ON PROCESS INTENSIFICATION THROUGH STORAGE REACTORS:A CASE STUDY ON METHANE STEAM REFORMING 68
Jack Lowd, Theodore T. Tsotsis, Vasilios Manousiouthakis

(568D) CO-PRODUCTION OF HYDROGEN AND BUTANOL BY FERMENTATION WITH THE MICROORGANISM CELLS IMMOBILIZED ON BAGASSE 69
Jingyun Liu, Zeyi Xiao, Senqing Fan

(568E) SOLAR THERMOCHEMICAL H2O/CO2 SPLITTING CYCLE USING FERRITES 72
Rahul Bhosale, Suliman Rashid

(623A) ADVANCED CARBON CAPTURE SYSTEM FOR POWER PLANTS 73
Gokhan Alptekin, Ambalavanan Jayaraman, David Gribble, Michael Bonnema, Chakravarthy Sishtla

(623D) THE USE OF DIGITAL TOOLS IN THE SEARCH FOR NEW SOLVENTS TO CAPTURE ACID GASES LIKE CO2 74
Frédérick De Meyer

(623E) COMPARATIVE STUDIES OF ENERGY SAVING PERFORMANCE FOR CO2 CAPTURE 83
Yuki Sato, Yasuki Kansha

(623F) DESIGN OF A FULLY INTEGRATED ARTIFICIAL PHOTOSYNTHETIC SYSTEM FOR A MOISTURE-GRADIENT CO2 CAPTURE AND REDUCTION 84
Aditya Prajapati, Meenesh R. Singh

(623G) STUDY ON CO CONVERSION FROM CO2 BY CHEMICAL LOOPING PROCESS 85
Kenji Kamiya, Nobusuke Kobayashi, Yoshinori Itaya, Tsuguhiko Nakagawa

(623H) CO2 SOLUBILITY IN WATER CONFINED BY KEROGEN NANOPORE IN RELATION TO CO2 SEQUESTRATION IN DEPLETED SHALE RESERVOIRS: A MOLECULAR DYNAMICS SIMULATION STUDY 87
Wenhui Li, Zhehui Jin

(626A) CHARACTERIZATION OF FACTORS AFFECTING THE GAS WELL DELIVERABILITY OF THE JIAOSHIBA SHALE GAS FIELD BY MULTIVARIATE STATISTICAL ANALYSIS 88
Yizhong Zhang, Maolin Zhang, Bin Ju, Fanhua Zeng, Haiyan Mei, Ping Guo

(626B) SOLUTIONS OF C6F13- AND C4F9-BASED POLYFLUOROACRYLATES IN CO2; SOLUBILITY, VISCOSITY, AND ADSORPTION ONTO CEMENT 110
Peter Lemaire III, Adel Alenzi, Peter Koronaios, Dustin Crandall

(626C) COMPETITIVE SORPTION OF CO2 IN SHALE NANOPORES FOR SEQUESTRATION AND ENHANCED GAS RECOVERY USING MOLECULAR DENSITY FUNCTIONAL THEORY 112
Jinlu Liu, Arjun Valiya Parambathu, Walter G. Chapman

(626D) SPATIAL AND TEMPORAL EVOLUTION OF NATURAL GAS AND NATURAL GAS LIQUID FEEDSTOCKS IN SHALE BASINS 113
Qining Chen, Jennifer B. Dunn, David Allen

(626E) A NEW GEOLOGY-ENGINEERING INTEGRATION SOLUTION FOR SHALE GAS EXPLORATION: A CASE STUDY FROM CHANGNING BLOCK IN CHINA 115
Jianfa Wu, Haoyong Huang, Cheng Chang, Yi Song, Shengxian Zhao

(626F) NEW TECHNOLOGY OF EOR OF ULTRA DEEP AND COMPLEX CONDENSATE GAS RESERVOIRS IN TARIM BASIN,CHINA 116
Jiang Tongwen, Xie Wei, Wang Yong

(626G) PRODUCTION CAPACITY EVALUATION METHOD OF CONDENSATE GAS RESERVOIR CONSIDERING THE INFLUENCE OF STRESS SENSITIVITY AND CONDENSATE OIL 117
Chunxiao Du, Fanhua Zeng, Jun Yang

(626H) NUMERICAL SIMULATION STUDY OF COMPLEX FRACTURE TYPE EXTRA HIGH CONDENSATE GAS RESERVOIR IN BOHAI SEA 124
Zhennan Gao, Yanchun Su, Zhigang Geng, Weilin Liu, Baogang Huang, Fanhua Zeng, Jun Yang

(698A) CHARACTERISTICS OF PHASE CHANGING MATERIAL NANOEMULSIONS 134
Artur Zych, Jungeun Park, Dinesh Kalaga, Robert Messinger, Masahiro Kawaji

(698B) MOLECULAR-LEVEL UNDERSTANDING OF PHASE BEHAVIOR IN PHASE-CHANGE NANO-EMULSIONS FOR THERMAL ENERGY STORAGE 135
Jungeun Park, Robert Messinger

(698C) EFFECTS OF OPERATING CONDITIONS FOR A SILICA GEL/WATER VAPOUR ADSORPTION THERMAL ENERGY STORAGE SYSTEM 136
Strong Curtis, Ye Carrier, F. Handan Tezel

(698D) EXPERIMENTAL INVESTIGATION OF A HIGH-TEMPERATURE THERMOCHEMICAL STORAGE REACTOR 137
Michael Wild, Lorenz Lüönd, Aldo Steinfeld

Author Index

601857 Modeling the Reactions and Transport in Gas Diffusion Electrode (GDE) for the Electrochemical Reduction of Nitrogen to Ammonia

Abstract:

Electrochemical reduction of N_2 to NH_3 is a promising alternative to the conventional Haber-Bosch process as the later is highly energy intensive and causes excess CO_2 emissions. The electrochemical synthesis of ammonia suffers from two major setbacks: 1) mass transfer limitations due to low solubility of N_2 in aqueous medium 2) Competing H_2 evolution reaction. The former can be overcome by depositing the catalyst in a porous Gas Diffusion Electrode (GDE). [1, 2] GDE has two parts, the catalyst layer which is hydrophilic to water and the gas diffusion layer which is hydrophobic to water. N_2 enters through the gas diffusion layer and reaches the catalyst surface. It has been experimentally observed that the current density of ammonia is two orders of magnitude higher than the planar electrode when GDE is used.

The improvement in current density is due to the higher surface area, the suppression of mass transfer limitation of N_2 to reach the catalyst surface, and minimum concentration polarization near the cathode surface. In this work we developed a multi-physics continuum model for the entire reactor system, that solves the transport of N_2 and water across the GDE, transport of N_2 in the electrolyte, charge transport and the kinetics of the Nitrogen Reduction Reaction (NRR) and the competing HER (Hydrogen Evolution Reaction). The model has been compared with the experimental results. The effect of various operating conditions such as pH, flow rates of N_2 and electrolytes has been studied. The effect of GDE parameters such as porosity, thickness of the catalyst and permeability has been investigated. Different reactor configurations have been simulated at various operating conditions to improve the existing performance of the catalyst.

References

1. Weng, L.C., A.T. Bell, and A.Z. Weber, *Modeling gas-diffusion electrodes for CO2 reduction.* Phys Chem Chem Phys, 2018. **20**(25): p. 16973-16984.
2. Higgins, D., et al., *Gas-Diffusion Electrodes for Carbon Dioxide Reduction: A New Paradigm.* ACS Energy Letters, 2018. **4**(1): p. 317-324.

609786 Method of Moments to Characterize Porous Media Using Non-Ideal Tracer

Abstract:

Tracer tests are widely used to characterize the porous media such as chemical reactors, oil reservoirs, groundwater aquifers. Due to solid phase heterogeneity, tracers may behave non-ideally and can get adsorbed on the solid surface. This leads to a long tail in tracer exit age density curves. In addition to the non-ideality of the tracer, the dead-ends in the porous media also lead to long tails. Simulations are used to estimate the dead-end volume fraction of the pores, mass transfer rate into the dead end, dispersion and adsorption behavior using the tracer exit age density curves. However, the parameter matching is difficult as the problem to find process parameters given the exit age density is ill-posed and can have multiple solutions. In this work, we presents a method of moments to characterize the porous medium for the process parameters using a single exit age density curve using a non-ideal adsorbing tracer. We use behavior of higher order moments of exit age density, namely, mean, variance, skewness, and kurtosis to estimate Peclet number, dead end volume fraction, mass transfer in dead end volume and adsorption behavior of the tracer. We observe that the first moment depends only on the retardation factor, which is always greater than unity for the non-ideal tracer. Similarly, we find the affect of the dead-end volume and adsorption of tracer on the variance, skewness, and kurtosis of the residence time density curve. We use experimental results from literature to establish the usability of our method.

600338 Evolution and Interaction Characteristics of Liquid Flow and Bubbles in a Jet Bubbling Column

Abstract:

Jet bubbling column reactors are commonly used in carbonylation to produce chemicals, such as butadiene carbonylation to adipic acid, butene carbonylation to valeraldehyde, and carbonylation to produce methyl methacrylate, etc. The jet bubbling column is a new type of gas-liquid two-phase reactor with excellent mixing ability. It is essential to understand the flow and interaction characteristics of the gas and liquid phases, to improve the performance of a jet bubbling column and to provide operation guidance for industrial reactors.

The evolution and interaction characteristics of liquid flow and bubbles in a jet bubbling column are numerically studied through a Computational Fluid Dynamic-Population Balance Model (CFD-PBM) coupling method. The axial velocity distribution of liquid velocity at different heights in the column is measured experimentally and compared with the simulation results to verify the accuracy of the CFD-PBM coupling model.

Simulation results show that the flow field in the column varies periodically. The period under the jet bubbling condition is longer than that under the pure jet condition. As shown in Figure 1, during a period, the liquid vortices continuously entrain and release bubbles, which can be observed experimentally. The liquid has both transporting and breaking effects on bubbles. On the one hand, the liquid vortices that move periodically transport the bubbles to various regions of the column, which is beneficial to the gas-liquid effective mixing in the column; on the other hand, the strong turbulence at the exit of the jet nozzle promotes the breakup of the bubbles. As the height increases, the proportion of bubbles with a diameter of smaller than 1 mm gradually increases, and the proportion of bubbles with a diameter of larger than 1 mm gradually decreases, as shown in Figure 2. The main region for bubble breakup is between the exit of the jet nozzle and 100 mm below it. We have proven that the CFD-PBM coupling model employed can be used to simulate the flow field of a jet bubbling column, which is of great significance for the industrial application and optimization of the jet bubbling column reactors.

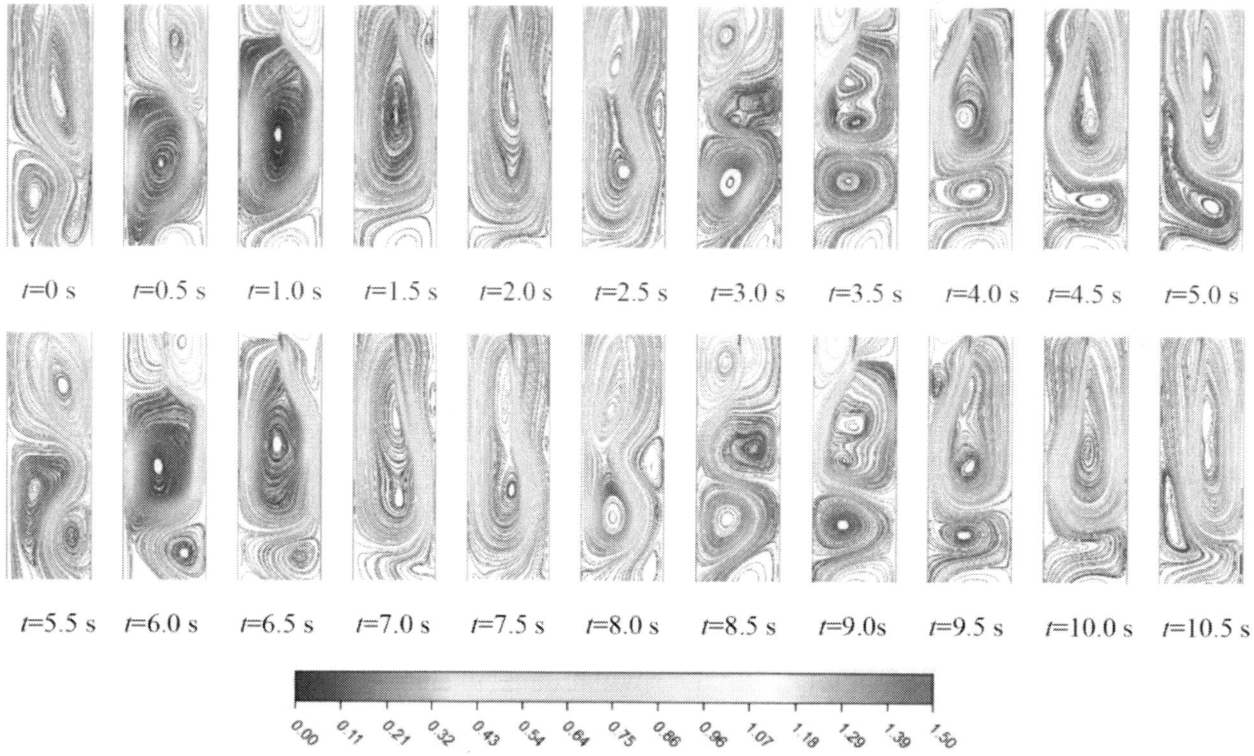

Figure 1 Liquid streamlines under the pure jet condition in a period (u_j=4.22 m/s)

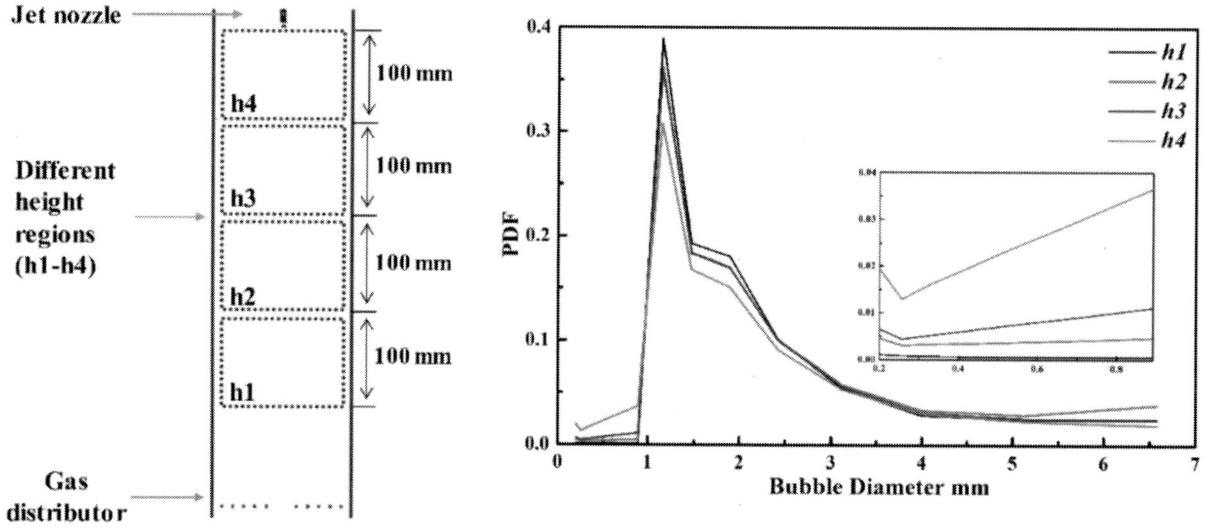

Figure 2 Bubble size distribution at different height regions

605016 Rational Design of Solid-Liquid Interphases for High-Energy Sodium Metal Batteries

Abstract:

Advances in the basic science and engineering principles of electrochemical energy storage is imperative for significant progress in electronic devices and grid stations. Metal based batteries comprising of a reactive metal (like Li, Na, Al) as the anode have attracted remarkable attention due to their promise of improving the anode-specific capacity by as much as 10-fold, compared to the current state-of-art Li-ion battery that uses a graphitic anode. A persistent challenge with batteries based on the metallic anode, concerns their propensity to fail due to short-circuits produced by dendrite growth during battery recharge, as well as by runaway of the cell resistance due to internal side reactions with the liquid electrolyte. In this talk, I will discuss my research that utilizes ion transport modeling and contemporary experimental efforts to fundamentally understand and to thereby develop rational designs for electrode-electrolyte interphases that overcome these challenges for sodium metal anodes. On the basis of a linear stability analysis of dendrite growth during metal electrodeposition, we have showed that a small fraction of immobilized anions near the electrodes is important in stabilizing metals against dendrite formation. To evaluate this proposal, we designed polymeric electrolytes with tunable molecular weight and quantified the stability of metal electrodeposition in these systems. Direct visualization of electrodeposition using these electrolytes showed remarkable agreement with the theoretical predictions. Furthermore, when operated in a battery, the polymeric electrolytes demonstrated stable and dendrite-free galvanostatic polarization of sodium metal metals for over 100 hours at relatively high current density.

605319 Quantification of Lithium Plating and Sources of Capacity Loss during Fast Charging of Graphite Electrodes

Abstract:

Fast charging of lithium-ion batteries is often hindered by rapid capacity fade, which is spurred by Li metal deposition, or 'Li plating,' on the graphite anode. Plated Li is problematic for multiple reasons, including the low reversibility of the Li plating/stripping process and the highly reactive nature of Li metal with electrolyte. Understanding when Li plating occurs and how much Li has plated is therefore crucial for improving battery lifetime. To this end, we demonstrate a titration technique which is used to quantify inactive Li and solid electrolyte interphase (SEI) species (solid carbonates and lithium acetylide, Li_2C_2) that remain on the graphite electrode after fast charging. We also employ electrochemical modeling in conjunction with experimental data to determine the Li plating exchange current density (10 A/m^2) and stripping efficiency (65%) of plated Li metal on graphite. In this work, we provide quantitative insights into the limitations of fast charging and determine valuable kinetic parameters for the Li plating process.

608173 Lithium Plating Detection during Extreme Fast Charging in Lithium-Ion Batteries Using Simultaneous Neutron and X-Ray Imaging

Abstract:

Lithium-ion batteries (LIBs) have profoundly advanced the development of electric vehicles (EV). However, one of the remaining bottlenecks in the widespread deployment of EVs is the long charging time required for these commercial LIBs. There is a global push towards extreme fast charging (XFC) to reduce their charging times to 10-15 minutes[1]. However, XFC results in severe degradation in the electrochemical performance of batteries. This is mainly attributed to the loss of active Li, either as "dead" which has become electronically disconnected after plating on the anode or as "inactive" due to the irreversible reaction of Li with the electrolyte to form a solid electrolyte interphase. Since parasitic lithium plating is one of the identified major results of XFCBs[2], understanding its origin and characteristics is important to developing strategies to enable fast-charged batteries. Several studies have characterized Li plating during XFC using techniques such as cryo-electron microscopy, transmission electron microscopy, X-ray diffraction, and X-ray micro-computed tomography (CT) among others[3]. While valuable, these characterization techniques have limitations[4] in terms of their sample preparation, imaging resolution, and the image contrast required to distinguish between materials of similar atomic number (Z).

Here, we are using simultaneous neutron and X-ray-based dual-mode micro-computed tomography, an alternative, non-destructive imaging modality, to investigate the characteristics of Li plating during XFC in LIBs. Since X-rays and neutrons are sensitive to the electron and nuclear density of the material respectively, dual-mode X-ray and neutron CT offers advantageous to readily separate the different anode components in a LIB such as Li, graphite, and pores due to the complementary interaction of the two imaging probes with matter. Higher energy X-rays are needed to get through the metallic components in batteries such as copper and aluminum. However, they do not have sufficient contrast to separate lithium from graphite. Thus, neutrons are used due to their high sensitivity to lighter elements such as lithium, especially ^{6}Li, and carbon.

We performed the multi-modal imaging experiments at the Neutron and X-ray Tomography (NeXT) [5] system located on the BT-2 beamline at National Institute of Standards and Technology Center for Neutron Research. We characterized pristine and cycled graphite anode strips containing plated lithium at different regions. For cycled anode strips, we disassembled the battery pouch cells and used the graphite anodes cycled under XFC conditions for 450 cycles at fully discharged condition. These strips were used to get sample sizes small enough to achieve the highest possible spatial resolution. In addition, we characterized an uncycled single-layer battery pouch cell containing electrolyte for which we rolled the battery cell in the form of a cylinder to put in the beam path. The spatial resolution achieved by this technique was ~10-15 μm, thus providing sufficient resolution to pinpoint the location of Li plating within the thickness of the anode. We are using image processing algorithms and bivariate histogram segmentation to process these dual-imaging datasets to investigate whether a link exists between the spatial heterogeneity of Li plating and the local microstructure of the battery anode such as porosity, tortuosity, and thickness. We are interested in understanding why and where Li plating occurs on the battery anode. Addressing these fundamental questions will inform improved battery designs, graphite anode architectures, and charging protocols that will reduce Li plating during

XFC in LIBs. We envisage that these findings will help make charging an EV similar to filling up at a gas station.

References:

1. Liu, Y., Zhu, Y., & Cui, Y. (2019). Challenges and opportunities towards fast-charging battery materials. Nature Energy, 1.
2. Tomaszewska, A., Chu, Z., Feng, X., O'Kane, S., Liu, X., Chen, J., ... & Li, Y. (2019). Lithium-ion battery fast charging: A review. eTransportation, 1, 100011.
3. Lu, J., Wu, T., & Amine, K. (2017). State-of-the-art characterization techniques for advanced lithium-ion batteries. Nature Energy, 2(3), 17011.
4. Li, Y., Li, Y., & Cui, Y. (2018). Catalyst: how cryo-EM shapes the development of next-generation batteries. Chem, 4(10), 2250-2252.
5. LaManna, J. M., Hussey, D. S., Baltic, E., & Jacobson, D. L. (2017). Neutron and X-ray Tomography (NeXT) system for simultaneous, dual modality tomography. Review of Scientific Instruments, 88(11), 113702.

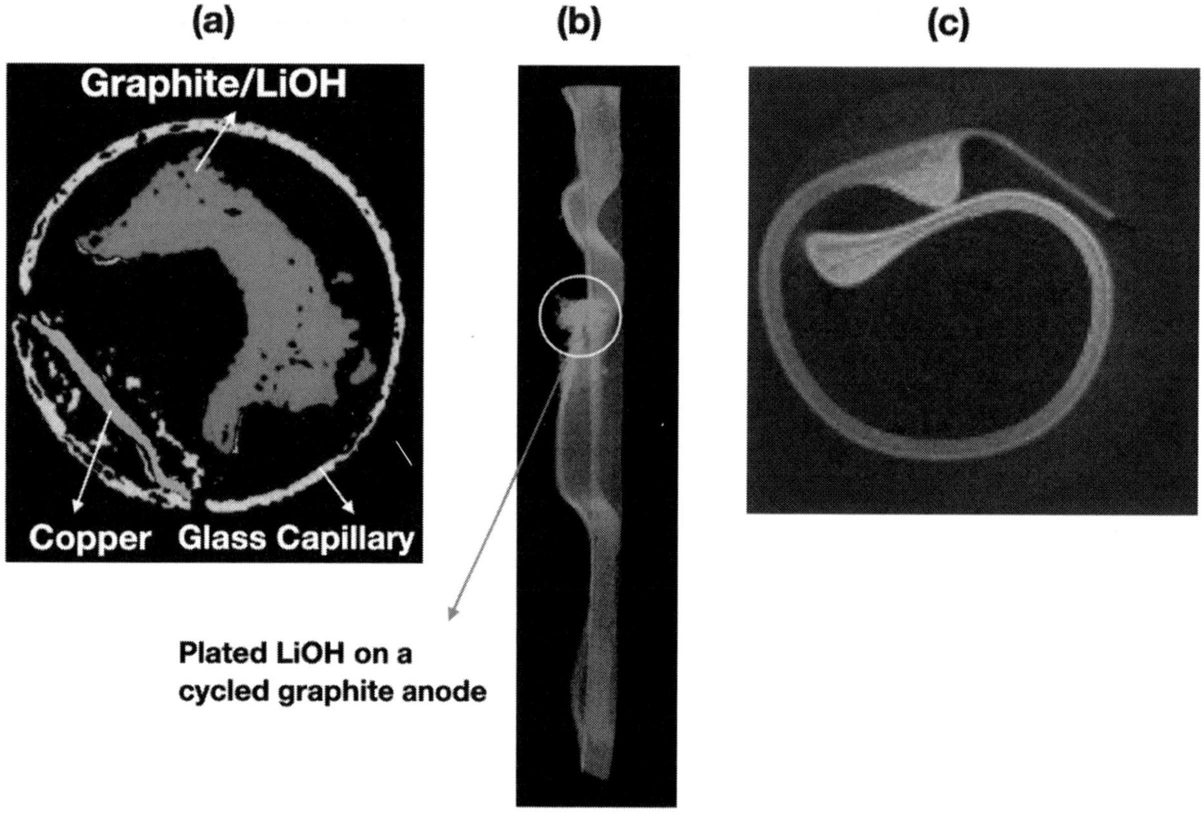

Figure 1. (a) *Top-View of the segmented image of the graphite anode containing plated LiOH in a glass capillary.* ***(b)*** *Dual tomography 3D reconstruction. Brown = 10 μm copper, pink = 80 μm graphite/ LiOH.* ***(c)*** *Neutron reconstruction of the single-layer battery pouch cell.*

611101 Surface Degradation of Li-Ion Battery Cathodes from First Principles

Abstract:

Nickel-rich high-energy-density Li-ion battery cathodes are prone to thermal degradation, which leads to safety concerns and hinders commercial adoption. Degradation typically occurs at the particle surface and involves the release of oxygen gas. To identify stable Ni-rich cathode compositions or surface coatings that can prevent undesired phase transformation, an understanding of the fundamental degradation processes on the atomic scale is needed.

Here, we apply first principles atomistic modeling to shed light on surface reactivity, providing information that is complementary to experimental characterization. The mechanism of the self-reductive decomposition of $LiNiO_2$ is determined using automated density-functional theory calculations. We construct first principles phase diagrams that map out stable surface phases as a function of the state of charge and the temperature. The results obtained from this analysis allow deducing the degradation mechanism and offer guidance for the development of degradation-stable cathode compositions and improved coating materials.

611581 Understanding Electrochemical Reaction Mechanisms and Properties of Rechargeable Aluminum-Sulfur and Aluminum-Selenium Batteries

Abstract:

Rechargeable Al metal batteries have recently garnered significant interest due to its low cost, earth abundance, inherent safety, and high volumetric (8040 mAh cm $^{-3}$) and gravimetric (2980 mAh g $^{-1}$) capacities due to the trivalent nature of Al^{3+} ions (3-electron redox). Commercialization of rechargeable aluminum battery systems has been limited, however, due to the small number of (i) electrolytes that enable reversible electrodeposition of Al metal at room temperature and (ii) positive electrode materials that are compatible with the electrolytes while providing high capacity and cycle life. Sulfur (S) and selenium (Se) are promising positive electrode materials for rechargeable battery systems as they exhibit very high specific capacities during electrochemical conversion reactions. To date, there are few reports of Al-S, and Al-Se battery systems. There is no clear consensus on the electrochemical reaction mechanism during the charge/discharge process, while capacity fade plagues both systems during galvanostatic cycling.

Here, aluminum-sulfur (Al-S) and aluminum-selenium (Al-Se) cells were investigated with a chloroaluminate ionic liquid electrolyte, $AlCl_3$:[EMIm][Cl] (molar ratio of 1.5:1), whose electrochemical reaction mechanisms and ensuing reaction products were analyzed via a combination of bulk (XRD, NMR) and surface (XPS) analyzation techniques.For Al-S batteries, electrochemical experiments showed high initial capacity, where both capacity and cycle life exhibited a non-monotonic relationship with the galvanostatic cycling rate. A higher capacity was observed upon charge, compared to discharge, pointing towards unwanted side reactions and/or polysulfide-shuttle-like behavior. For the Al-Se batteries, the two different reaction mechanisms that have been previously reported were found to be dependent on (i) the applied current density and (ii) the crystallinity of the active material. We demonstrate that, with appropriate cycling conditions, both reaction mechanisms can occur during a single charge/discharge step, thus significantly increasing the capacity and the electrochemical window of the Al-Se cell.

To understand the reaction mechanisms of Al-S and Al-Se batteries at a molecular level, solid-state and liquid-state ^{27}Al NMR and ^{77}Se NMR measurements (on Al-Se systems) were acquired on the electrodes and electrolyte at different states-of-charge, respectively. XPS was also performed on cycled chalcogen electrodes to understand surface compositions. The data yield insights into the reaction mechanisms and products, including aluminum coordination environments and the quantitative relative populations of aluminum in these environments. Overall, this work establishes the potential of conversion-type aluminum-chalcogen batteries as high-energy-density energy storage systems and highlights the importance of molecular-level analytical tools like NMR spectroscopy in clarifying reaction mechanisms in emerging electrochemical systems.

591791 Electrochemical Kinetics of Degradation of Porous Graphite Electrodes in Lithium Ion Batteries

Abstract:

Growth of the solid electrolyte interphase (SEI) is a major driver of capacity fade in LIBs. Despite its importance, the fundamental mechanisms remain unclear, primarily because of the complicated reaction pathways involved[1–3]. SEI growth can be both electrochemical and chemical in nature[4], and thus, it is a strong function of the potential and degree of lithiation of the electrode. We model the early-stage and long-term growth of SEI by accurately capturing the potential dependence of its formation kinetics as well as long term rate limiting steps, and validating it against the world's largest open source battery cycling data, generated in-house[5]. This is done using the Multiphase Porous Electrode Theory (MPET) framework[6] on graphite (phase separating) and carbon black (non phase separating) particles.

Lithium plating is another key degradation phenomenon that has been elusive, and it becomes important while trying to fast-charge batteries, i.e., 0% - 80% state-of-charge in 30 mins. We show that lithium plating is a key function of electrode morphology, phase-separation dynamics and potential. Phase-separation in graphite is modeled in the electrode using the Cahn-Hilliard Reaction framework described by Bazant[7]. We understand the electrochemistry of the onset of lithium plating with in-situ measurements connected to real time cell potential in a phase-separating electrode for the first time[8].

Results indicate that the peak SEI-forming currents are higher for higher driving currents. Also, we find that SEI only grows during electrode lithiation, i.e. the battery only degrades while being charged. We also find that onset of lithium plating is correctly captured only when phase separation in active material is accounted for. Further, the onset of plating is delayed on electrodes with a thick SEI layer – understanding SEI/plating coupling is integral to predicting fast charging manufacturing protocols for LIBs. This work holds promise for the predictive design of procedures[9] for manufacture and formation of LIBs.

[1] Cohen, Y. S.; Cohen, Y.; Aurbach, D. Micromorphological Studies of Lithium Electrodes in Alkyl Carbonate Solutions Using in Situ Atomic Force Microscopy. J. Phys. Chem. B 2000, 104, 12282–12291. https://doi.org/10.1021/jp002526b.
[2] Horstmann, B.; Single, F.; Latz, A. Review on Multi-Scale Models of Solid-Electrolyte Interphase Formation, Current Opinion in Electrochemistry 13, 62-69 2019.
[3] Nie, M.; Abraham, D. P.; Seo, D. M.; Chen, Y.; Bose, A.; Lucht, B. L. Role of Solution Structure in Solid Electrolyte Interphase Formation on Graphite with LiPF6 in Propylene Carbonate. J. Phys. Chem. C 2013, 117 (48), 25381–25389, https://doi.org/10.1021/jp409765w.
[4] Das, S.; Attia, P. M.; Chueh, W. C.; Bazant, M. Z. Electrochemical Kinetics of SEI Growth on Carbon Black: Part II. Modeling. J. Electrochem. Soc. 2019, 166 (4), E107–E118. https://doi.org/10.1149/2.0241904jes.
[5] Attia, P. M.; Das, S.; Harris, S. J.; Bazant, M. Z.; Chueh, W. C. Electrochemical Kinetics of SEI Growth on Carbon Black: Part I. Experiments. J. Electrochem. Soc. 2019, 166 (4), E97–E106. https://doi.org/10.1149/2.0231904jes

[6] Smith, R. B.; Bazant, M. Z. Multiphase Porous Electrode Theory, J. Electrochem. Soc. 2017, 164 (11). https://doi.org/10.1149/2.0171711jes

[7] Bazant, M. Z., Theory of chemical kinetics and charge transfer based on nonequilibrium thermodynamics, *Accounts of Chemical Research*, *46*(5), 1144–1160. https://doi.org/10.1021/ar300145c

[8] T. Gao, Y. Han, S. Das, T. Zhou, D. Fraggedakis, N. Nadkarni, C. N. Yeh, W. Chueh, J. Li, M.Z. Bazant, Interplay of lithium intercalation and plating on graphite using in-situ optical measurements, submitted.

[9] Huang, W.; Attia, P. M.; Wang, H.; Renfrew, S. E.; Jin, N.; Das, S.; Zhang, Z.; Boyle, D. T.; Li, Y.; Bazant, M. Z.; McCloskey, B. D.; Chueh, W. C. and Cui, Y.; Nano Letters 2019 19 (8), 5140-5148. DOI: 10.1021/acs.nanolett.9b01515

595921 Strategies for Performance Improvement of Polyoxometalates in Non-Aqueous Redox Flow Batteries

Abstract:

Polyoxometalates have been explored as multi-electron active species in non-aqueous redox flow batteries. Although non-aqueous systems offer a wider voltage window for redox flow battery operation, realization of this potential requires a judicious choice of solvent as well as polyoxometalate properties. We demonstrate here the superior performance of N,N-dimethylformamide (DMF) compared to acetonitrile as a solvent for redox flow batteries based on $Li_3PMo_{12}O_{40}$. This compound displays two 1-electron transfers in acetonitrile but can access an extra quasi-reversible 2-electron redox process in DMF. A cell containing 10 mM solution of $Li_3PMo_{12}O_{40}$ in DMF produced a cell voltage of 0.7 V with 2-electron transfers (State of Charge = 60%) and showed a good cyclability. As a strategy to boost energy density, operation of the redox flow battery at a higher concentration of 0.1 M $Li_3PMo_{12}O_{40}$ produced cells with cell voltage of 0.6 V in acetonitrile and cell voltage of 1.0 V in DMF; both showed excellent coulombic efficiencies of more than 90% over the course of 30 cycles. Another strategy to improve the performance is by employing an asymmetric cell with different polyoxometalates on each side to extend cell voltage. $Li_6P_2W_{18}O_{62}$ exhibited 3 quasi-reversible 2-electron transfers in the potential range between -2.05 V to -0.5 V vs. Ag/Ag^+. 10 mM $Li_6P_2W_{18}O_{62}/Li_3PMo_{12}O_{40}$ in DMF produced a cell with cell voltage of 1.3 V involving 4-electron transfers (State of Charge = 50%) with coulombic efficiency of nearly 100% and energy efficiency of nearly 70% throughout the test with more than 20 cycles. These promising results demonstrate proof-of-concept approaches to improving the performance of polyoxometalates in non-aqueous redox flow batteries.

596974 Bridging Mechanism of V^{2+}/V^{3+} Reaction on Glassy Carbon in Acidic Electrolytes for Vanadium Redox Flow Batteries

Abstract:

With the world's energy demand expected to increase immensely in coming decades, along with the adverse effects of CO_2 emissions on environment by burning of fossil fuels, there is a need to shift to renewable energy sources like solar and wind. These technologies are expected to generate ~50% of total world's electricity by 2050, almost double to the current levels. However, despite the enormous potential of renewable resources in solving the world's electricity demands, they suffer from intermittency necessitating storing the excess electricity and releasing based on demand. Although pumped hydroelectric currently provides the majority of electricity storage, its geographical limitations prevent its further scaling, motivating searches for alternative electricity storage technologies.

Redox Flow Batteries (RFBs) have shown tremendous capability to store and release renewable electricity at a multimegawatt-hours scale with long electrode lifetimes. RFBs have a unique ability to decouple energy and power which provides them high scalability and flexibility for grid-energy storage applications. Among aqueous RFBs, all-vanadium RFBs (VRFBs: $VO_2^+/VO^{2+}//V^{2+}/V^{3+}$) are among the most developed RFBs and the closest to commercial implementation. However, VRFBs suffer from a high kinetic overvoltage of the slower V^{2+}/V^{3+} redox couple ($V^{2+} \rightleftarrows V^{3+} + e^-$, V vs SHE) that lowers the overall battery efficiency.[1] Despite substantial work on VRFBs, there is a lack of fundamental understanding of V^{2+}/V^{3+} redox couple chemistry, particularly the role of electrolyte and the associated charge transfer. VRFBs with carbon electrodes employing hydrochloric acid (HCl) or mixed acid electrolytes (HCl/H_2SO_4)[1] have higher current densities compared with the traditionally used sulfuric acid (H_2SO_4) electrolyte, but it is unclear if the enhancements are due to kinetics, mass transport, or conductivity.

Recently, we have reported V^{2+}/V^{3+} kinetic measurements on glassy carbon in HCl, H_2SO_4, and mixed HCl/H_2SO_4 electrolytes for VRFBs. We extract exchange current densities (i_o) via steady-state current measurements ($i_{o,Tafel}$) and impedance spectroscopy ($i_{o,Rct}$) and apparent activation energies (E_a). Our results from both independent techniques to measure i_o show that the V^{2+}/V^{3+} reaction kinetics in HCl are ~2.5 times faster than in H_2SO_4 or HCl/H_2SO_4. By comparing values of $i_{o,Rct}$ to $i_{o,Tafel}$, we also show that V^{2+}/V^{3+} reaction is an overall two electron transfer reaction in HCl (with one electron in rate determining step, RDS), as opposed to an overall one electron transfer reaction in H_2SO_4 and HCl/H_2SO_4. We propose that the enhanced V^{2+}/V^{3+} redox kinetics in HCl are caused by the charge transfer proceeding through more polarizable chloride (*Cl) bridge, instead of through surface-bound hydroxyl (*OH) groups in H_2SO_4 and HCl/H_2SO_4.[2] We further hypothesize that the anion's bridging strength, which is a measure of deformability in the presence of electric field,[3] will govern the V^{2+}/V^{3+} reaction kinetics, with stronger bridging strength corresponding to higher activity.

In this talk, we will discuss our V^{2+}/V^{3+} kinetic measurements on glassy carbon in hydrobromic (HBr) and hydriodic (HI) acid to test the hypothesis of bridging strength as the sole criteria governing V^{2+}/V^{3+} reaction kinetics. The bridging strength of anions follows the order, I > Br > Cl > OH,[3] so based on our hypothesis we expect the V^{2+}/V^{3+} reaction rates to follow the same trend. The V^{2+}/V^{3+} exchange current densities in HBr are ~1.5 times higher than in HCl, with 5 to 15 kJ mol^{-1} lower E_a (Figure 1). The i_o increases and the E_a decreases with increase in State of Charge (SoC) in HBr, similar to the behavior observed in HCl and H_2SO_4. We determine that the reaction in HBr is an overall two electron transfer reaction, with one electron involved in RDS, similar to HCl. Further, the V^{2+}/V^{3+} reaction is HI has kinetics faster than HBr, with a maximum rate at ~50% SoC. The $i_{o,Rct}$ in HI indicates the reaction mechanism changes to an overall one electron transfer. However, we notice that the rates in HI are affected immensely by the presence of small amount of sulfate anions. This points to the bridging strength not being a sole descriptor for the reaction kinetics. We attribute this observed trend in rates of V^{2+}/V^{3+} reaction to be dependent on the stability of the vanadium intermediate formed on the surface of glassy carbon, computed using density functional theory.

REFERENCES:

1. Choi, C. *et al. Renew. Sustain. Energy Rev.* **69**, 263–274 (2017).
2. Agarwal, H. *et al. ACS Energy Lett.* **4**, 2368–2377 (2019).
3. Anbar, M. *et al. J. Phys. Chem.* **69**, 973–977 (1965).

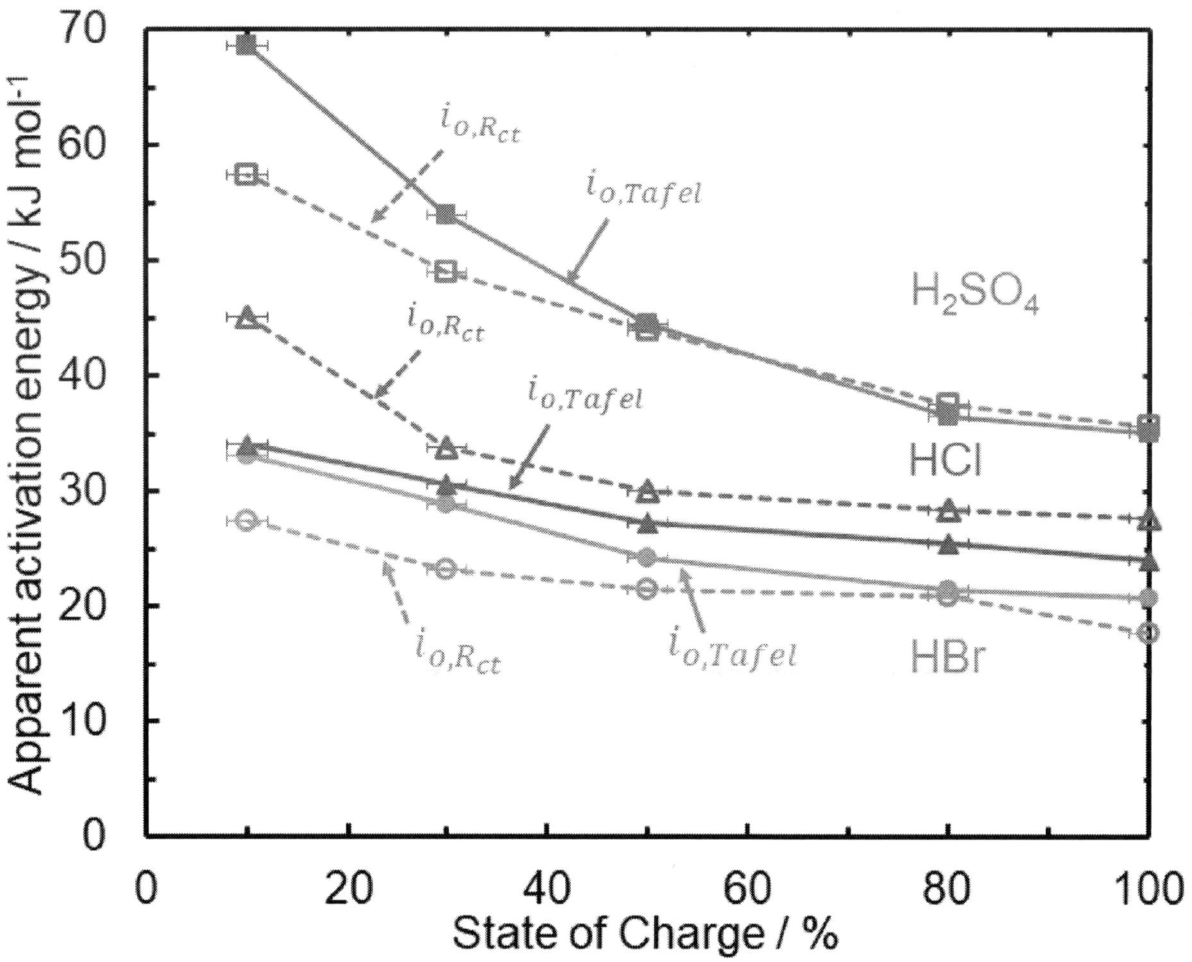

Figure 1. Apparent activation energies (E_a) for the V^{2+}/V^{3+} redox reaction on a glassy carbon working electrode evaluated for different states of charge (SoC) in H_2SO_4 (0.2 M $VOSO_4$ + 0.5 M H_2SO_4), HCl (0.2 M VCl_3 + 1 M HCl), and HBr (0.1 M V_2O_5 + 1 M HBr) electrolytes from exchange current densities obtained by two independent methods: a) (solid line) By using Tafel equation ($\boldsymbol{i_{o,Tafel}}$), and b) (dashed line) Charge transfer resistance ($i_{o,R_{ct}}$) at open circuit voltage using potentiostatic electrochemical impedance measurements. The $i_{o,Tafel}$ and $i_{o,R_{ct}}$ were evaluated at each SoC for four different temperatures: 23.3, 30.0, 35.0, and 40.0 °C. Counter electrode is graphite rod separated from working electrode side by Nafion N117 membrane. Reference electrode is Ag/AgCl. Apparent activation energies in H_2SO_4 and HCl are reproduced from ref. 2.

601387 Electrochemical Systems Optimization Using Physics-Based Homogenization of the Microscopic Continuum Models

Abstract:

Electrochemical systems such as fuel cells, flow batteries, and electrochemical reactors are prevalent in industry, but few tools for their numerical design and optimization exist. Simulation of these systems spans length-scales from sub-micron to meter and often requires the solution of multiple coupled partial differential equations (e.g. mass transport, momentum transport, and electrostatics). Appropriate resolution of the typically porous, electroactive component domain is often computationally expensive, making shape and topology optimization at this scale especially difficult if not prohibitive. To accelerate and enable the design optimization of these systems, we employ physics-based homogenization of the governing microscopic, continuum transport equations. In this presentation, we briefly review the homogenization formulation for porous electrochemical systems and our numerical methodology for extracting effective properties. We then apply these techniques to develop an up-scaled representation for a flow battery electrode composed of an architected porous medium consisting of an iso-truss lattice. Through resolved computations on the microstructural unit cell, we parameterize its macroscopic response as a function of the unit cell porosity to generate the homogenized model. By applying the adjoint method, we calculate model sensitivities with respect to the spatially varying of unit cell porosities, enabling design optimization at the device scale. We conclude by applying our methodology to generate optimized architectures for scale-up of flow-through electrodes.

This work was performed under the auspices of the U.S. Department of Energy by Lawrence Livermore National Laboratory under Contract DE-AC52-07NA27344.

594685 CFD Simulations of Tangentially Fired Coal Boiler Under Various Loads

Abstract:

Tangentially fired boilers are widely used in coal power plants. The pulverized coal particles and air form a fireball in the center of boiler with intense concentric swirling turbulent mixing. The nation's generating assets are changing as more electricity being generated from intermittent renewable capacity such as wind and solar power plant. Coal-fired power plants optimized as baseload resources are being increasingly operated in partial load situations. Existing coal power plants are being operated in ways that are suboptimal from the perspective of efficiency and capacity utilization. In an effort to address these challenges by identifying and developing technologies that increase coal-fired power plant efficiency, improve unit reliability and availability, and enhance unit capability for flexible operations (e.g. "cycling"), a CFD model using ANSYS-FLUENT is developed to simulate the industrial scale coal boiler.

The Discrete Phase Models is used to track particles. The Discrete Ordinates (DO) radiation model, realizable k-ε turbulence model, gas phase chemical reaction, coal devolatilization, char combustion are included in the heat transfer. Eddy-dissipation is used as the turbulence-chemistry interaction model. The CFD model is first validated with the available data which had been collected from Tri-State's Escalante coal power plant. After the CFD model is validated with experimental results, then a much wider range of operational parameters such as the pulverize air, secondary air and coal particle flowrates are simulated and their effect on combustion, temperature distribution, flow field are analyzed and discussed.

602606 Multiobjective Control of Fiber Morphology in Pulping Process Via Multiscale Modeling

Abstract:

It is well known that the mechanical properties of paper products are mainly dependent upon the morphological characteristics of wood pulp due to the interfacial adhesion of individual fibers [1]. Particularly, cell wall thickness and length of fibers are important parameters as they are closely related to the tensile and burst strengths of final paper products [1-2]. Despite the importance of fiber morphology, the regulation of fiber morphology to desired values during pulping process has not been enabled because of two challenges. First, even though wood fibers undergo broad morphological changes under conditions of Kraft pulping, online measurements of the properties are unavailable due to the harsh operating conditions such as high cooking temperature and corrosive liquor. Second, given the increasing PPI's (Pulp and Paper Industry) reliance on model-based technology to optimally adjust the process while conforming the strengthened environmental regulation, several kinetic models have been developed to describe the kinetics and transport behavior of the Kraft pulping process [3-9]. Nevertheless, as the models are derived from macroscopic equations (i.e., differential equations) that are proven to fully capture the bulk phase dynamics, they come with an inbuilt weakness; elementary mechanisms that occur on very small scales such as degradation of cell wall and fiber breakage are hardly described by them.

Motivated by the limitations, in this work, a multiscale model is developed to simulate the evolution of fiber morphology in a pulp digester. Specifically, by integrating the most widely used mathematical model for pulping process (i.e., the Purdue model) with a kinetic Monte Carlo (kMC) algorithm, the evolution of both the Kappa number (i.e., residual lignin content in pulps), and fiber morphology (i.e., cell wall thickness and fiber length) are accurately captured; degradation events of cell wall components are executed based on the kMC algorithm, and the pit distribution and fiber breakage probability model are utilized to compute the fiber length. Then, a reduced-order model is identified using the high-fidelity input/output data from the proposed multiscale model to handle the computational requirement of the developed model [10]. While incorporating the developed multiscale model into a model-based controller design, an interesting process control problem is found; particularly, simultaneously driving both the Kappa number and fiber length is infeasible as they are two conflicting objectives. In order to accomplish the low Kappa number which is typically required from most of paper grades, it is unavoidable to produce fibers with thin cell wall thickness, which in turn leads to fiber breakage. Therefore, we employed the epsilon-constraint method to find the Pareto optimal solutions which allow decision makers to choose a preferred operating condition depending on a target product quality. Based on the optimal solutions, a model-based controller is designed to achieve them while minimizing the heat usage during Kraft pulping.

References

[1] Dinwoodie JM. The relation between fiber morphology and paper properties. TAPPI J. 1965;48(8):440-447.

[2] Xu F.; Zhong X.; Sun R.; Lu Q.; Jones GL. Chemical composition fiber morphology, and pulping of P. bolleana lauche. Wood Fiber Sci. 2006; 38(3), 512-519

[3] Wisnewski PA.; Doyle FJ. Model-based predictive control studies for a continuous pulp digester,

IEEE Transactions on control systems technology, 2001; 9(3), 435-444.

[4] Facada MJ. Influence of Kraft paper quality on the performance of an industrial paper impregnation process. Ph.D. thesis, Universidade Tecnica de Lisboa, Lisboa, Portugal, 2015.

[5] Funkquist J. Grey-box identification of a continuous digester – a distributed-parameter process. Control Eng. Pract., 1997, 5, 919-930.

[6] Wisnewski PA.; Doyle FJ. Fundamental continuous pulp-digester model for simulation and control. AIChE J., 1997, 43, 3175-3192.

[7] Smith C. Studies of the mathematical modeling simulation and control of the operation of Kamyr continuous digester for the Kraft process. Ph.D. thesis, Purdue University, West Lafayette, IN, 1974.

[8] Bhartiya S.; Dufour P.; Doyle FJ. Fundamental thermal-hydraulic pulp digester model with grade transition. AIChE J., 2003, 49, 411-425.

[9] Choi HK.; Kwon JSI. Modeling and control of cell wall thickness in batch delignification. Comput. Chem. Eng. 2019, 128, 512-523.

[10] Chou CT.; Verhaegen M. Subspace algorithms for the identification of multivariable dynamic errors-in-variables model. Automatica, 1997, 33, 1857-1869.

605175 Weak Formulation of Supercooled Stefan Problem for Solidification of Pure Silicon in Horizontal Ribbon Growth

Abstract:

Solar energy is a promising source of energy due to its abundance, and has been backed by technological advancements, lower cost and rising global demand. Despite this the silicon wafer costs are still considered high and accounts for 40% of the photo-voltaic supply chain [1]. This is in part because current ingot based wafer production techniques like Czochralski and Bridgman processes---which contribute more than 80% of the silicon substrate for solar cells---are batch processes with low production rates (~2 mm/min) and are highly wasteful [2]. These processes produce up to 55% of material loss while sawing followed by tedious, time consuming grinding and lapping operations. This can be significantly reduced if the wafers could be directly solidified to its correct thickness [3].

Silicon, like water, has a lower density in its solid form than its liquid form. This property of silicon can be used to solidify single-crystal silicon on top of its melt to form a thin sheet of ribbon [4]. The ribbon can then be continuously extracted horizontally by cooling the top surface of the melt. This process, descriptively called Horizontal Ribbon Growth (HRG), has the potential for significant cost savings over traditional wafer production methods [1]. Because the wafers must only be cut to length, there is a substantial decrease in material loss. Additionally in the HRG process, the crystal grows epitaxially in the direction orthogonal to the direction of the heat sink. This makes it possible to produce thin single crystal ribbons at a relatively high speed [5].

Despite its advantages, the HRG process faces fundamental challenges which inhibit its successful operation. One of these is the theoretical modeling of the solidification and heat transfer process around the growing tip of the ribbon. Traditional heat transfer models have been unsuccessful in predicting the upper-limit on the pull speed of the ribbons [6,7]. Finite element simulations of the Stefan problems predict that it is possible to grow a crystal ribbon as thin as desired by increasing the pull speed of the ribbon without limit [9]. However, experimental observations have shown that this is not possible and a very distinct upper bound to the pull speed exists.

Recent numerical models have attributed the pull-speed limit to the non-existence of a stable meniscus or to the formation of a {111} facet at the leading edge [8,9]. To-date no prediction or relationship has been proposed to model the upper bound on pull speed. We identify the reason for this pull speed disparity to be a non-differentiability condition at the tip of the growing ribbon. We conclude that the irregularity of the solid-liquid interface inhibits the use of the classical Stefan formulation to predict the dynamics of the solidifying tip.

To circumvent this problem, we propose to use a weak formulation of the Stefan problem at the solidification boundary. The notion of a sharp interface separating the solid and liquid domains is replaced by a mushy region of a liquid and solid mixture. The phenomenon of supercooling, which is generally observed in the solidification of a pure substance, is modeled using a nucleation dependent phase-temperature rule. This restricts phase change to occur only around a small neighborhood of an existing solid. The proposed model is thus aimed at simulating evolution of a growing solid conditioned on the availability of a nearby nucleating site. The model enjoys the simplicity of the weak formulation, while at the same time incorporating the metastability of the supercooled liquid.

The underlying theory for our heat transfer solution of solidification is reviewed and numerical simulations are presented which highlight the relationship between pull-speed and heat transfer. A finite volume approach was used to study the heat transfer for a variety of boundary conditions including radiation, cooling plate, and helium jets. In each of these cases, an upper limit to the pull speed is observed. The upper limit to the pull speed is shown to be a fundamental consequence of the heat removed from the tip of the ribbon. Recommendations to enhance production speeds and control thickness of the ribbon are also proposed. While our application study focuses on the HRG process, the theory we develop can be applied to other solidification problems as well.

References:

[1] Ranjan, S., Balaji, S., Panella, R. A., & Ydstie, B. E. (2011). Silicon solar cell production. Computers & Chemical Engineering, 35(8), 1439-1453.

[2] Oliveros, G. A., Liu, R., Sridhar, S., & Ydstie, B. E. (2013). Silicon wafers for solar cells by horizontal ribbon growth. Industrial & Engineering Chemistry Research, 52(9), 3239-3246.

[3] William, S. (1962). U.S. Patent No. 3,031,275. Washington, DC: U.S. Patent and Trademark Office.

[4] Eyan P Noronha, German A Oliveros, and Erik B Ydstie. Weierstrass' variational theory

for the stability of meniscus in ribbon growth processes, under review.

[5] Ke, J., Khair, A. S., & Ydstie, B. E. (2017). The effects of impurity on the stability of Horizontal Ribbon Growth. Journal of Crystal Growth, 480, 34-42.

[6] Zoutendyk, John A. "Theoretical analysis of heat flow in horizontal ribbon growth from a melt." Journal of Applied Physics 49.7 (1978): 3927-3932.

[7] Zoutendyk, John A. "Analysis of forced convection heat flow effects in horizontal ribbon growth from the melt." Journal of Crystal Growth 50.1 (1980): 83-93.

[6] Daggolu, Parthiv, et al. "Thermal-capillary analysis of the horizontal ribbon growth of silicon crystals." Journal of Crystal Growth 355.1 (2012): 129-139.

[8] Helenbrook, Brian T., et al. "Experimental and numerical investigation of the horizontal ribbon growth process." Journal of Crystal Growth 453 (2016): 163-172.

605732 Designing and Optimizing Energy Policies through a Novel Energy Price Predictive Framework

Abstract:

Energy prices in many countries do not reflect environmental damage, notably climate change, air pollution and various health related side effects, so targeted fiscal policies should be designed and applied to address this issue [1]. In addition, there is an imbalance in the countries' fiscal objectives with excessive emphasis being put on general income, payroll and consumption taxes and limited emphasis on use of energy taxes [2]. Therefore, a tax reform towards energy-smarter designed taxes is required to mitigate climate change while raising well-being, having positive effects on poverty and equity, and enabling countries and firms to become more climate resilient and productive[3]. Consequently, fiscal and monetary policies shall be leveraged to drive climate action. To this respect, it becomes crucial to accurately quantify the "price of energy" and understand the effects that fiscal and monetary policies will have in energy.

To address both challenges, we developed a novel predictive framework, the Texas A&M Energy Price Index (EPIC) that can be used as a benchmark to calculate the average price of energy to the end-use consumers in the United States, as well as an effective tool to assess, design and optimize various policy rules and questions [4]. Here, we demonstrate the effectiveness of this framework by addressing two contemporary policy questions that have raised substantial public and governmental interest. Specifically, the effects of a gasoline tax hike on EPIC are examined parametrically considering various scenarios i.e. 5, 10, 15, 20 and 25 cents per gallon increase in federal gasoline taxes. Moreover, the effects of renewable energy production targets and subsidies on energy consumers are examined parametrically over a wide range of different weights of the energy feedstocks as well as for tax credits ranging from 0 to 9 US$/MMBtu. For both case studies, the different scenarios are presented both for the past i.e. what would have happened, as well as for the future i.e. what will happen, along with the revenue generated or the budget required for the implementation of each scenario, taking advantage of the powerful predictive ability and flexibility of the proposed framework.

References:

1. OECD (2019), *Taxing Energy Use 2019: Using Taxes for Climate Action*, OECD Publishing, Paris, https://doi.org/10.1787/058ca239-en.
2. W. Parry, M. D. Heine, E. Lis, and S. Li, Getting energy prices right: From principle to practice. International Monetary Fund, 2014.
3. A. Pigato, Fiscal Policies for Development and Climate Action. The World Bank, 2018.
4. Baratsas, S.; Niziolek, A.; Onel, O.; Matthews, L.; Floudas, C. A.; Hallermann, D.; Sorescu, S.; Pistikopoulos, E. N. What is the price of energy to the end-users? A novel predictive framework and one of its applications to the energy and monetary policies. Nature Communications 2020 (Under Review).

607599 Identifying Optimal Kinetic Pathways of a Functionalized Solid Sorbent for CO_2 Capture through Mathematical Programming

Abstract:

Recent research has been directed into finding viable solid sorbents for post-combustion CO_2 capture that can be alternatives to aqueous solvents which are the current leading commercial technology. For these sorbents to be attractive on the commercial scale, they must exhibit lower regeneration energy, faster kinetics, higher CO_2 selectivity, and higher working capacities. To realize the full potential of these systems, accurate predictive models that can be used for system design is absolutely necessary. Specifically, accurate modeling of the adsorption equilibrium for CO_2 as well as the other components present in a flue gas source is a very important and crucial step in developing these full system models. Current modeling for most sorbents uses classical isotherm models, such as a Langmuir isotherm model, to predict the adsorption equilibrium. However, when they are functionalized and complicated chemisorption interactions arise, these simple models fail to accurately represent the equilibrium. Traditional isotherm models have been developed for physical adsorption. These models fail to represent the complex equilibrium behavior of reactive adsorbents. While there has been attempts in the literature on developing weighted isotherm models, these models are highly parameterized. These models are neither predictive, nor provide any insight into the reaction pathways and products formed. A fundamental understanding of the reaction pathways can be instrumental in designing the functionalized MOFs for higher loading, better kinetics and lower energetics.

The sorbent studied in this work is a novel functionalized metal-organic framework (MOF) that has been shown to exhibit many of the important characteristics for CO_2 capture[1,2,3]. The MOF, Mg_2(dobpdc) (dobpdc^{4-} = 4,4′-dioxidobiphenyl-3,3′-dicarboxylate), contains one-dimensional hexagonal channels with unsaturated Mg^{2+} sites and is functionalized by binding the diamine 2,2-dimethyl-1,3-diaminopropane (dmpn) to these sites which then chemisorb CO_2 in cooperative mechanism to form carbamate and carbamic acid species. This cooperative adsorption results in step shaped isotherms which are very difficult to predict using traditional isotherm models. Also, experimental work has been done to investigate which products are formed in the presence of CO_2 but the dominating reactions and pathways that give these products are still relatively unknown[3]. Since the reactions take place at functionalized sites in the MOF and exhibit the step-shaped behavior, it is impractical to experimentally determine reaction products and reaction pathways as a function of partial pressure of CO_2 and temperature. To determine which set of reactions can best represent the system chemistry, first a set of plausible reactions including chain formation reactions and numerous ways that these chains can interact are proposed based on the information from the sparse NMR data. A mixed integer non-linear programming (MINLP) problem is solved to identify the optimal kinetic pathway as well as model parameters considering both physisorption and chemisorption. Since growing chain lengths can lead to a highly parameterized system, an information-theoretic criterion is considered as the optimization objective. For thermodynamic consistency, energetics of the candidate reactions are considered as constraints. The results of this model development are then compared to previous attempts at modeling the adsorption equilibrium using traditional methods.

References:

[1] – Milner, P.J., Siegelman, R.L., Forse, A.C., Gonzalez, M.I., Runcevski, T., Martell, J.D., Reimer, J.A., Long, J.R. A Diaminopropane-Appended Metal-Organic Framework Enabling Efficient CO_2 Capture from Coal Flue Gas via a Mixed Adsorption Mechanism. *Journal of the American Chemical Society.* 2017; 139 (38), 13541-13553

[2] Mcdonald, T.M., et al., Cooperative insertion of CO_2 in diamine-appended metal-organic frameworks. *Nature.* 2015; 519, 303-308

[3] Forse, A.C., Milner, P.J., Lee, J., Redfearn, H.N., Oktawiec, J., Siegelman, R.L., Martell, J.D., Dinakar, B., Porter-Zasada, L.B., Gonzalez, M.I., Neaton, J.B., Long, J.R., Reimer, J.A., Elucidating CO_2 Chemisorption in Diamine-Appended Metal-Organic Frameworks. *J. Am. Chem. Soc.* 2018; 140(51), 18016-18031

Received: 28 June 2020 | Revised: 2 September 2020 | Accepted: 5 September 2020
DOI: 10.1002/aic.17052

PROCESS SYSTEMS ENGINEERING

AIChE JOURNAL

Equilibrium analysis of CH_4, CO, CO_2, H_2O, H_2, C mixtures in C–H–O atom space using Gibbs free energy global minimization

Demetrios Chaconas[1] | Patricia Pichardo[1] | Ioannis V. Manousiouthakis[2] | Vasilios I. Manousiouthakis[1]

[1]Department of Chemical and Biomolecular Engineering, Hydrogen Engineering Research Consortium (HERC), University of California at Los Angeles (UCLA), Los Angeles, California
[2]Hydrogen Engineering Research Company (H-E-R-C), Los Angeles, California

Correspondence
Vasilios I. Manousiouthakis, Department of Chemical and Biomolecular Engineering, Hydrogen Engineering Research Consortium (HERC), University of California at Los Angeles (UCLA), Los Angeles, CA 90095.
Email: vasilios@ucla.edu

Abstract

Gibbs free energy minimization is employed to carry out thermodynamic equilibrium analysis studies of mixtures containing methane, carbon monoxide, carbon dioxide, water, and hydrogen ideal gases, and possibly solid carbon. The employed global minimization approach represents a general, unifying, conceptual framework that allows reaction and phase equilibrium analysis to be simultaneously carried out in the atom-mol fraction space (a_H, a_O), thus capturing in a comprehensive manner the equilibrium behavior of a number of industrially important processes, such as methane reforming (steam, dry, energetically enhanced), and methanation. Two theorems are presented, establishing necessary and sufficient (necessary) feasibility and regularity (optimality) conditions for the aforementioned minimization problem. The equilibrium results obtained through application of these theorems are guaranteed to be globally optimal. They quantify in (a_H, a_O) space, the feasible region, the carbon formation region, and all species mole over total atom-mol normalized ratios, for a range of temperatures and pressures.

KEYWORDS
optimization, equilibrium, Gibbs, thermodynamics/classical

1 | INTRODUCTION

The identification of fundamental performance limitations of chemical processes, by carrying out chemical and phase equilibrium studies, is a time-honored tradition that helps accelerate the development of commercially successful industrial processes. Chemical equilibrium studies have been traditionally carried out using equilibrium constant based methods. A set of reversible chemical reactions is first contemplated, which subsequently gives rise to a number of nonlinear algebraic equations relating each reaction's equilibrium constant with the participating species equilibrium fugacities (compositions for ideal mixtures). The species' feed compositions are needed to carry out the relevant computations. Gibbs' pioneering work established the conceptual foundation for simultaneous chemical and phase equilibrium studies by proving that at constant pressure and temperature, a system reaches chemical and phase equilibrium if and only if the system's total Gibbs free energy is minimized (infimized).[1] In this approach, no reversible chemical reactions need to be contemplated, and only a list of participating species, and the atom-moles of participating elements must be known.[1]

The focus of this work is the simultaneous reaction and phase equilibrium analysis of systems containing CH_4, CO, CO_2, H_2O, H_2, and possibly solid carbon C, as such systems are of great practical importance, since mixtures containing these species arise in numerous industrial processes, including steam methane reforming (SMR), energetically enhanced steam methane reforming (EESMR), dry methane reforming (DMR), and syngas methanation (SM). Formation of carbon (or coking) while carrying out these processes has been shown to have deleterious effects: it can cause catalyst deactivation, physical obstruction of process equipment, and the destruction of unit operations.[2] Thus, knowledge of regions in process parameter space (temperature, pressure, feed composition) in which carbon formation occurs, is of great importance so as to limit the aforementioned detrimental effects on industrial processes.

Steam methane reforming (SMR) is used on an industrial scale to generate about 95% of the world's hydrogen from CH_4 and H_2O. SMR is typically fed with H_2O:CH_4 molar ratios of around three,[3] so as to maximize hydrogen yield and prevent carbon formation. The latter has been the subject of many SMR equilibrium studies, several of which were reviewed in.[2] Considering that SMR equilibrium can be captured by the Methanation, Shift, and Boudouard reactions for systems containing CH_4, CO_2, CO, H_2O, H_2, and possibly C, the carbon formation boundary on a C, H, O, atom-mole fraction diagram was computed at 723 K, 0.14 MPa, using the equilibrium constant method.[2] Later, several carbon formation boundaries were produced at different temperature and pressure combinations considering a set of seven reactions representing steam methane reforming along with carbon-forming reactions.[4]

EESMR was put forward as a means of reducing the amount of CH_4 consumed as fuel in carrying out the endothermic SMR process, by adding CO and excess H_2O in the reformer feed.[5-9] The EESMR process was patented in 2017.[5] Subsequently, Pichardo and Manousiouthakis employed EESMR, atom balances, and Gibbs free energy minimization based equilibrium reactor models, within the Infinite-DimEnsionAl State-space (IDEAS) conceptual framework, to reduce by over an order of magnitude the hot utility consumption of methane reforming, compared to traditional SMR designs.[6] EEMSR equilibrium computations were carried out for a variety of CO/CH_4 and H_2O/CH_4 feed ratios, demonstrating that EESMR could deliver reformers with over 95% CH_4 conversions, and exothermic heat loads, thus enabling the creation of natural gas-based hydrogen production systems utilizing hybrid energy sources, such as natural gas and concentrated solar power (CSP).[7] Gibbs free energy minimization based equilibrium calculations were used to assess the execution of EESMR in membrane reactors, again demonstrating high CH_4 conversions, and exothermic heat loads, with even lower CO/CH_4 and H_2O/CH_4 feed ratios than those used in EESMR reactors.[8] Gibbs free energy minimization based EESMR reactor models were employed within a heat integrated, CH_4/CSP utilizing flowsheet, and EESMR was shown to be preferable in the presence of significant levels of taxation on the use of natural gas as fuel.[9]

DMR utilizes CO_2 as the oxidizing agent to convert CH_4 into syngas (mixture of CO and H_2), which is a precursor for the production of H_2 and organic chemicals (e.g., through the Fischer-Tropch process). DMR is an economically appealing and environmentally friendly process, whose commercialization is hindered by carbon formation.[10] For this reason, previous DMR research efforts have used the FactSage software to carry out thermodynamic equilibrium analysis to study the effect of pressure, temperature, and CH_4/CO_2 molar ratios on carbon formation.[11] Carbon deposition in the DMR process has previously been investigated and the results indicated that DMR should be carried out at temperatures greater than 1,000°C, a pressure of 0.1 MPa, and a CH_4/CO_2 molar feed ratio of 1 to ensure no carbon deposition.[11] H_2 is produced in the DMR process at H_2:CO molar ratios of one or less.[11]

In addition to reforming, syngas methanation (SM), that is, the conversion of syngas into methane, is another widely explored process. SM is commonly employed as the final purification step in the production of hydrogen, by which carbon oxides are removed from the product hydrogen stream.[12] Methanation is also being considered for the production of substitute natural gas (biogas) from syngas (biomass derived syngas), although these processes have been rendered economically infeasible as a wealth of natural gas resources have been discovered in recent years.[12] Carbon formation in CO methanation, for various feedstock compositions, temperatures, and pressures, was investigated experimentally,[13] using molecular orbital theory,[14] and through use of the equilibrium constant method applied to eight reactions involved in modeling methanation.[15] This latter study found that adding H_2O (CH_4) into the SM feed reduces (increases) carbon deposition, and methanation is favored by large hydrogen to carbon feed ratios, low temperatures, and high pressures.[15] In addition to Gibbs free energy minimization, equilibrium gas phase compositions were also determined via parametric equations, in the temperature range of 500–700 K, indicating that higher hydrogen to carbon feed ratios resulted in higher CO_2 conversion, with the best conversions under coking-free conditions occurring at ratios of 4:1 H_2:CO_2 or larger.[15]

The Gibbs free energy minimization and equilibrium constant methods have both been used to solve the chemical equilibrium problem for multi-component, ideal gas mixtures in the presence of adsorbents.[16] No heterogeneous reactions or condensed surface species were considered, but rather a Langmuir isotherm was employed to relate adsorbed and gaseous phase species amounts.[16] Computational challenges related to singularities near zero species concentrations were discussed, and the Lagrange multiplier method was used to transform the optimization problem into a set of nonlinear algebraic equations solved locally using the Newton–Raphson method, for an ethanol steam reforming case study, involving eight gaseous species and CO_2 adsorption.[16] In a later work, solid carbon formation and CO_2 adsorption was considered in the context of maximizing H_2 yields in ethanol reforming.[17] Parametric analysis was performed by varying temperature, pressure, and the steam-to-carbon ratio to identify the most promising operating conditions to produce H_2 and prevent coking.[17]

The above suggest that equilibrium studies for the aforementioned industrial processes could benefit from the development of a general unifying framework that would allow the computation of equilibrium compositions, feasibility regions, and carbon formation regions in process parameter space. The Gibbs free energy minimization method naturally lends itself for such a development, as it only requires knowledge of temperature, pressure, the species involved in the equilibrium, and the amounts of each element present in the feed, and requires no knowledge of reactions in which the species participate. Further, since the number of elements is typically lower than the number of species, the dimensionality of the parameter space for the resulting mathematical formulation is significantly reduced.[18]

In this work, we provide a general conceptual framework, based on Gibbs free energy minimization, that can be used to carry out simultaneously and in a unified manner, chemical-phase equilibrium studies for several industrial chemical processes. The paper is structured as follows. First, the Gibbs free energy based minimization problem is formulated in atom fraction space based on fundamental principles of thermodynamics. Then, a feasibility analysis of the minimization problem is carried out, conditions for the minimum's existence are developed, and the necessary conditions for optimality are derived. The global minimum for the optimization problem yielding the normalized equilibrium flowrates of each species is computed for a range of common industrial process temperatures and pressures, and for feed compositions spanning the entire feasible region in atom fraction space. Next, the carbon formation boundary is computed for a range of pressures and temperatures in atom fraction space, and lastly, conclusions are drawn.

2 | MATHEMATICAL FORMULATION

Gibbs' pioneering work,[1] established that, at constant pressure and temperature, a system reaches chemical and phase equilibrium if and only if the system's total Gibbs free energy is minimized (infimized). This work applies this principle in identifying the equilibrium behavior of a mixture consisting of the six substances, CO, CO_2, CH_4, H_2, H_2O, C. To streamline the presentation, consider the optimization (infimization) problem:

$$\left\{\begin{array}{l} \nu \triangleq \inf\limits_{\{x_j\}_{j=1}^{n} \in \mathbb{R}^n} f\left(\{x_j\}_{j=1}^{n}\right) \\ \text{subject to } h_i\left(\{x_j\}_{j=1}^{n}\right) = 0 \ \forall i = 1,m \\ g_k\left(\{x_j\}_{j=1}^{n}\right) \leq 0 \ \forall k = 1,p \end{array}\right\} \tag{1}$$

If the feasible region of ν is a non-empty, closed, and bounded set, and its objective function is continuous, then the existence of a minimum for ν is guaranteed.[18] Further, if all functions involved in defining the objective function and constraints of ν are differentiable, then at any candidate local minimum of ν that is a regular point, the following necessary optimality conditions must hold[19]:

$$\left\{\begin{array}{l} \dfrac{\partial f\left(\{x_j\}_{j=1}^{n}\right)}{\partial x_l} + \sum\limits_{i=1}^{m} \lambda_i \dfrac{\partial h_i\left(\{x_j\}_{j=1}^{n}\right)}{\partial x_l} + \sum\limits_{k=1}^{p} \mu_k \dfrac{\partial g_k\left(\{x_j\}_{j=1}^{n}\right)}{\partial x_l} = 0 \ \forall l = 1,n \\ h_i\left(\{x_j\}_{j=1}^{n}\right) = 0 \ \forall i = 1,m; \ \mu_k g_k\left(\{x_j\}_{j=1}^{n}\right) = 0 \ \forall k = 1,p \\ g_k\left(\{x_j\}_{j=1}^{n}\right) \leq 0 \ \forall k = 1,p; \ \mu_k \geq 0 \quad \forall k = 1,p \end{array}\right\} \tag{2}$$

where regularity is formally defined as:

$$\begin{array}{l} \{x_j\}_{j=1}^{n} \in \left\{h_i\left(\{x_j\}_{j=1}^{n}\right) = 0 \forall i = 1,m; g_k\left(\{x_j\}_{j=1}^{n}\right) \leq 0 \forall k = 1,p\right\} \text{ is regular} \Leftrightarrow \\ \left\{\left\{\dfrac{\partial h_i\left(\{x_j\}_{j=1}^{n}\right)}{\partial x_l}\right\}_{l=1}^{n} \forall i = 1,m; \left\{\dfrac{\partial g_k\left(\{x_j\}_{j=1}^{n}\right)}{\partial x_l}\right\}_{l=1}^{n} \forall k \in \left\{k = 1,p : g_k\left(\{x_j\}_{j=1}^{n}\right) = 0\right\}\right\} \text{ are linearly independent} \end{array} \tag{3}$$

Application of the aforementioned general optimization principles requires that the total Gibbs free energy minimization problem be first formally defined. At any fixed temperature–pressure pair within the ranges considered in the work, the studied species mixture is ideal, and its gaseous components are ideal gases. Then, the Gibbs free energy infimization problem is:

$$\left\{\begin{array}{l} \bar{\nu} \triangleq \inf\limits_{n_i\, i \in S} G(T,P,\{n_i\}_{i \in S}) \triangleq \sum\limits_{i \in S-\{C\}} \left[n_i \left(G_i^{\circ}(T) + RT \ln\left(\dfrac{n_i P}{\sum\limits_{k \in S-\{C\}} n_k}\right)\right)\right] + n_C\left[G_C^{\circ}(T)\right] \\ \text{subject to } \bar{a}_j - \sum\limits_{i \in S} \nu_{ij} \bar{n}_i = 0 j \in \{C,H,O\}; \ \bar{n}_i \geq 0 \ \forall i \in S \end{array}\right\} \tag{4}$$

where $G_i^{\circ}(T)$ (J/mol$i$) is the standard Gibbs free energy of the ith species at the temperature T(K) (whose evaluation utilizes data from Reference 20, and is summarized in Supporting Information, Appendix A.1), P(dimensionless) is the operating pressure normalized over the standard pressure of 0.1 MPa, $\bar{n}_i$(mol$i$) is the number of moles of species $i \in S \triangleq \{CO, CO_2, CH_4, H_2, H_2O, C\}$ in the system, ν_{ij} is the number of atoms of element j in species i, and $\bar{a}_j$ is the number of atom-moles of element j in the system. The objective function of (4) is the total Gibbs free energy $G(T,P,\{\bar{n}_i\}_{i \in S})$ (J), while its equality constraints represent atom balances, and its inequality constraints ensure that the number of moles of each participating species is nonnegative. Given that, for fixed temperature and pressure, the objective function of (4) is a 1st order homogeneous function, scaling the problem's parameters $\bar{a}_j$, $j \in \{C,H,O\}$ and variables $\bar{n}_i$, $i \in S$ by a scaling factor $\bar{a}$ (say

$\bar{a} \triangleq \sum_{j\in\{C,H,O\}} a_j$), then yields an equivalent problem whose optimum value, variables and parameters satisfy $\bar{\nu} = \bar{a}\cdot\nu$, $\bar{n}_i = \bar{a}\cdot n_i$, $i\in S$, and $\bar{a}_j = \bar{a}\cdot a_j$, $j\in\{C,H,O\}$, respectively. Further, close examination of (4) also suggests that its objective function is not a continuous and differentiable function (in the traditional two-sided sense) in portions of the feasible region's boundary (e.g., where $\bar{n}_i = 0$ for some $i \in S$). Thus, a constant $\varepsilon > 0$ such that $\varepsilon \to 0^+$ is employed in the inequality constraints of (4), to help ascertain objective function differentiability. The resulting scaled, simultaneous chemical-phase equilibrium, normalized total Gibbs free energy infimization problem ν_ε, satisfying $\nu = \lim_{\varepsilon\to 0}\nu_\varepsilon$, is stated as follows:

where n_i, $i \in S$ denote the mole over total atom-mol normalized ratios of all considered species, a_j is the atom-mol fraction of element j in the system, and ν_ε is the optimum value of the total atom-molar normalized Gibbs free energy. It is straightforward to establish that if the feasible region of (5) is non-empty, it is a closed and bounded set, since all constraints are linear functions, and all variables satisfy $0 \le \bar{n}_i \le \max_{j\in\{C,H,O\}} \bar{a}_j \forall i\in S$. Since the objective function of ν_ε is also continuous throughout the feasible region, then the Weierstrass theorem,[18] establishes that if the feasible region of ν_ε is non-empty, the minimum of ν_ε exists. Variable substitution then brings ν_ε into the following equivalent form, featuring no equality constraints.

$$\left\{\begin{array}{l} \nu_\varepsilon \triangleq \inf\limits_{\substack{n_{CO},n_{CH_4},\\ n_{H_2O},n_{H_2},\\ n_{CO_2},n_C}} n_{CO}G^\circ_{CO}(T) + n_{CH_4}G^\circ_{CH_4}(T) + n_{H_2O}G^\circ_{H_2O}(T) + n_{H_2}G^\circ_{H_2}(T) + n_{CO_2}G^\circ_{CO_2}(T) + n_C G^\circ_C(T) + \\ + RT\left(\begin{array}{l} n_{CO}\ln\left(\dfrac{n_{CO}P}{n_{CO}+n_{CH_4}+n_{CO_2}+n_{H_2O}+n_{H_2}}\right) + n_{CH_4}\ln\left(\dfrac{n_{CH_4}P}{n_{CO}+n_{CH_4}+n_{CO_2}+n_{H_2O}+n_{H_2}}\right) + \\ + n_{CO_2}\ln\left(\dfrac{n_{CO_2}P}{n_{CO}+n_{CH_4}+n_{CO_2}+n_{H_2O}+n_{H_2}}\right) + n_{H_2O}\ln\left(\dfrac{n_{H_2O}P}{n_{CO}+n_{CH_4}+n_{CO_2}+n_{H_2O}+n_{H_2}}\right) + \\ + n_{H_2}\ln\left(\dfrac{n_{H_2}P}{n_{CO}+n_{CH_4}+n_{CO_2}+n_{H_2O}+n_{H_2}}\right) \end{array}\right) \\ \text{s.t. } a_C-(n_{CO}+n_{CH_4}+n_{CO_2}+n_C)=0, a_H-(4n_{CH_4}+2n_{H_2O}+2n_{H_2})=0, a_O-(n_{CO}+n_{H_2O}+2n_{CO_2})=0 \\ \varepsilon-n_{CO}\le 0, \varepsilon-n_{CH_4}\le 0, \varepsilon-n_{H_2O}\le 0, \varepsilon-n_{H_2}\le 0, \varepsilon-n_{CO_2}\le 0, -n_C\le 0, a_C+a_H+a_O=1 \end{array}\right\} \tag{5}$$

$$\left\{\begin{array}{l} n_{CO}=2a_C-a_O-2n_{CH_4}+n_{H_2O}-2n_C, n_{CO_2}=a_O-a_C+n_{CH_4}+n_C-n_{H_2O}, n_{H_2}=\frac{1}{2}a_H-2n_{CH_4}-n_{H_2O} \\ \left\{\begin{array}{l} \nu_\varepsilon \triangleq \inf\limits_{\substack{n_{CH_4},\\ n_{H_2O}\\ n_C}} (2a_C-a_O-2n_{CH_4}+n_{H_2O}-2n_C)G^\circ_{CO}(T) + n_{CH_4}G^\circ_{CH_4}(T) + n_{H_2O}G^\circ_{H_2O}(T) + \\ + \left(\frac{1}{2}a_H-2n_{CH_4}-n_{H_2O}\right)G^\circ_{H_2}(T) + (a_O-a_C+n_{CH_4}+n_C-n_{H_2O})G^\circ_{CO_2}(T) + n_C G^\circ_C(T) + \\ + RT\left(\begin{array}{l} (2a_C-a_O-2n_{CH_4}+n_{H_2O}-2n_C)\ln(2a_C-a_O-2n_{CH_4}+n_{H_2O}-2n_C) + n_{CH_4}\ln(n_{CH_4}) + \\ + (a_O-a_C+n_{CH_4}+n_C-n_{H_2O})\ln(a_O-a_C+n_{CH_4}+n_C-n_{H_2O}) + n_{H_2O}\ln(n_{H_2O}) + \\ + \left(\frac{1}{2}a_H-2n_{CH_4}-n_{H_2O}\right)\ln\left(\frac{1}{2}a_H-2n_{CH_4}-n_{H_2O}\right) + \left(a_C+\frac{1}{2}a_H-2n_{CH_4}-n_C\right)\ln(P) - \\ - \left(a_C+\frac{1}{2}a_H-2n_{CH_4}-n_C\right)\ln\left(a_C+\frac{1}{2}a_H-2n_{CH_4}-n_C\right) \end{array}\right) \\ \text{s.t. } \varepsilon-2a_C+a_O+2n_{CH_4}-n_{H_2O}+2n_C\le 0, \varepsilon-a_O+a_C-n_{CH_4}-n_C+n_{H_2O}\le 0, \\ \varepsilon-\frac{1}{2}a_H+2n_{CH_4}+n_{H_2O}\le 0, \varepsilon-n_{CH_4}\le 0, \varepsilon-n_{H_2O}\le 0, -n_C\le 0 \\ a_C+a_H+a_O=1 \end{array}\right. \end{array}\right\} \tag{6}$$

The following theorem first identifies necessary and sufficient conditions in $a_C - a_H - a_O$ space for the optimization problem ι_ε to be feasible. Then, it identifies conditions in $a_C - a_H - a_O$ space for the feasible points of the optimization problem ι_ε to be regular.

Theorem 1 ***Theorem 1a***. *The optimization problem ι_ε is feasible if and only if:*

$\{$a.$)\ 3\varepsilon \le a_C$, b.$)\ 8\varepsilon \le a_H$, c.$)\ 4\varepsilon \le a_O$, d.$)\ 3\varepsilon \le a_C - 1/2a_O + 1/4a_H$
e.$)\ 3\varepsilon \le a_C - 1/4a_O + 1/8a_H\}$

Theorem 1b. *Consider that the optimization problem ι_ε is feasible, and assume that $a_C > 3\varepsilon$, $a_H > 8\varepsilon$, $a_O > 4\varepsilon$. Then, all feasible points of ι_ε are regular, except for those satisfying the conditions below, which are not regular when the corresponding qualifying conditions hold:*

Proof 1 See Supporting Information, Appendix A.2.

Based on Theorem 1, and given that $a_C = 1 - a_H - a_O$, the feasible region of ι_ε, and the $a_H - a_O$ lines for which irregular points of ι_ε may exist, can be graphically illustrated in $a_H - a_O$ space, as shown in Figure 1, for $\varepsilon = 10^{-16}$. The feasible region's upper boundary is determined by the lines $0 = a_H + a_O - 1 + 3\varepsilon = -a_C + 3\varepsilon$, and $0 = 3/2a_O + 3/4a_H - 1 + 3\varepsilon$, which intersect at the point $(a_H, a_O) = ((2/3 - 2\varepsilon), (1/3 - \varepsilon))$, that also belongs to the line $0 = 5/4a_O + 7/8a_H - 1 + 3\varepsilon$, which everywhere else lies outside the feasible region. The physical interpretation of the feasible region being a strict subset of the triangle defined by the vertices (0, 0), (0, 1), (1, 0) is that the $(a_H, a_O) = (0, 1)$ vertex cannot be reached because all considered oxygen containing substances (CO, CO_2, H_2O) contain either hydrogen, or carbon atoms as well. The so-called candidate irregular point lines shown in Figure 1, identify points in $a_H - a_O$ space for which the corresponding feasible points $(n_{CH_4}, n_{CO}, n_{CO_2}, n_{H_2O}, n_{H_2}, n_C)$ identified in Theorem 1b are irregular. As will be shown in the Theorem 2, these irregular points must be separately considered as potential global optima, for (a_H, a_O) points belonging to the candidate irregular point lines.

Having identified the feasible region of ι_ε in $a_H - a_O$ space, next conditions are identified, for the existence of the global optimum of ι_ε, and for candidate global optima of ι_ε.

Theorem 2 *Let* $\mathrm{A}(T) \triangleq \frac{-2G_{CO}(T) + G_{CH_4}(T) - 2G_{H_2}(T) + G_{CO_2}(T)}{RT}$, $\mathrm{B}(T) \triangleq \frac{G_{CO}(T) + G_{H_2O}(T) - G_{H_2}(T) - G_{CO_2}(T)}{RT}$ $\Gamma(T) \triangleq \frac{-2G_{CO}(T) + G_{CO_2}(T) + G_C(T)}{RT}$, $\alpha(T) \triangleq \exp[\mathrm{A}(T)]$, $\beta(T) \triangleq \exp[\mathrm{B}(T)]$, $\gamma(T) \triangleq \exp[\Gamma(T)]$, *and assume* $\{3\varepsilon < a_C$, $8\varepsilon < a_H$, $4\varepsilon < a_O$, $3\varepsilon \le a_C - 1/2a_O + 1/4a_H$, and $3\varepsilon \le a_C - 1/4a_O + 1/8a_H\}$. *Then, the global minimum* $\{n_{CH_4}, n_{CO}, n_{CO_2}, n_{H_2O}, n_{H_2}, n_C\}$ *of the feasible optimization problem ι_ε exists, and is either one of the feasible and irregular points identified in Theorem 1b, or is a feasible, regular point that in conjunction with a six-plet* $\{M_{CH_4}, M_{CO}, M_{CO_2}, M_{H_2O}, M_{H_2}, M_C\}$ *satisfies the following necessary conditions for optimality:*

$$\left\{\begin{array}{l}
(1)'\ \frac{\beta(T)^2\gamma(T)^2}{\alpha(T)} = \left[\frac{(n_{CH_4})(2a_C - a_O - 2n_{CH_4} + n_{H_2O} - 2n_C)}{(n_{H_2O})^2}\right]\left(\frac{M_C{}^2 M_{H_2O}{}^2}{M_{CO_2}M_{CH_4}}\right) \\
(2)'\ P\frac{\gamma(T)}{\alpha(T)} = \left[\frac{(n_{CH_4})\left(a_C + \frac{1}{2}a_H - 2n_{CH_4} - n_C\right)}{\left(\frac{1}{2}a_H - 2n_{CH_4} - n_{H_2O}\right)^2}\right]\left(\frac{M_C M_{H_2}}{M_{CH_4}}\right) \\
(3)'\ \frac{\beta(T)\gamma(T)^2}{\alpha(T)} = \left[\frac{(n_{CH_4})(2a_C - a_O - 2n_{CH_4} + n_{H_2O} - 2n_C)}{(n_{H_2O})\left(\frac{1}{2}a_H - 2n_{CH_4} - n_{H_2O}\right)}\right]\left(\frac{M_C{}^2 M_{H_2O} M_{H_2}}{M_{CH_4}M_{CO}}\right) \\
(4)'\ n_{CO} = 2a_C - a_O - 2n_{CH_4} + n_{H_2O} - 2n_C,\ (5)'\ n_{CO_2} = a_O - a_C + n_{CH_4} + n_C - n_{H_2O} \\
(6)'\ n_{H_2} = \frac{1}{2}a_H - 2n_{CH_4} - n_{H_2O},\ (7)'\ (M_{CO} - 1)(\varepsilon - n_{CO}) = 0,\ (8)'\ (M_{CO_2} - 1)(\varepsilon - n_{CO_2}) = 0 \\
(9)'\ (M_{H_2} - 1)(\varepsilon - n_{H_2}) = 0,\ (10)'\ (M_{CH_4} - 1)(\varepsilon - n_{CH_4}) = 0, \\
(11)'\ (M_{H_2O} - 1)(\varepsilon - n_{H_2O}) = 0,\ (12)'\ (M_C - 1)(-n_C) = 0 \\
\varepsilon - n_{CO} \le 0,\ \varepsilon - n_{CO_2} \le 0,\ \varepsilon - n_{H_2} \le 0,\ \varepsilon - n_{CH_4} \le 0, \varepsilon - n_{H_2O} \le 0,\ -n_C \le 0 \\
1 - M_{CO} \le 0, 1 - M_{CO_2} \le 0, 1 - M_{H_2} \le 0, 1 - M_{CH_4} \le 0, 1 - M_{H_2O} \le 0, 1 - M_C \le 0 \\
n_{CO} + n_{CO_2} + n_{CH_4} + n_{H_2O} + n_{H_2} = a_C + \frac{1}{2}a_H - 2n_{CH_4} - n_C
\end{array}\right\} \qquad (7)$$

For a given (a_H, a_O), all solutions of the above set of equations and inequalities are identified by considering the cases tabulated in Table 1 below, and by solving, for each case, a polynomial equation in a single variable for all its roots. Then, the global optimum *of ν_ε is chosen as the solution that minimizes the objective function of ν_ε over the finite number of feasible, regular points resulting from the cases in the Table 1 below, and the finite number of feasible, irregular points identified in Theorem 1b, when applicable.*

Proof 2 See Supporting Information, Appendix A.3.

The global minimum of the objective function ν_ε is determined as a function of T, P, a_O, and a_H in the following manner. First, T and P are selected and fixed. Second, a value of a_H is fixed, and a range of a_O values are considered for the $\{T, P, a_H\}$ selected. The selected a_O values encompass the feasible and irregular points listed in Theorem 1b to ensure that these points are considered. For values of a_O in which all of the feasible points are regular, we compare only the candidates satisfying the necessary conditions of optimality for the objective function ν_ε. For the a_H, a_O combinations at which both regular and irregular points are feasible, we compute the value of the objective function for the irregular points, apply the necessary conditions of optimality for the regular points, and then select the smallest value from the two sets as the global optimum.

It is only for the regular points, at which the necessary conditions for optimality are applicable, that the equilibrium constant method of analysis holds, as it arises naturally from the case in which all species are present at equilibrium; that is, $n_i > \varepsilon \Rightarrow M_i = 1 \ \forall \ i = \{CH_4, CO, CO_2, H_2O, H_2\}$. For the irregular points, or when one or more species may not be present in amounts larger than infinitesimal at the optimum, the equilibrium constant method of analysis is not applicable. In turn, for this work, this implies that Theorem 2b must be employed to identify the globally optimal solution to ν_ε.

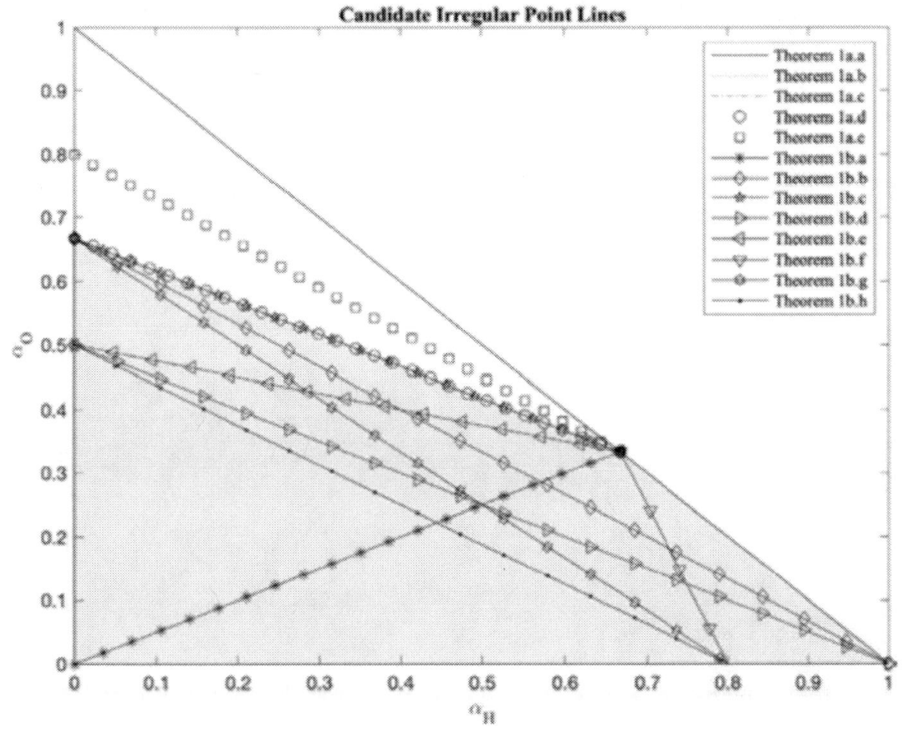

FIGURE 1 Feasible region (shaded) and candidate irregular points in $a_H - a_O$ space ($\varepsilon = 10^{-16}$)

TABLE 1 Feasible and irregular cases; (=) signifies $n_i = \varepsilon$

Irregular case	n_{CH_4}	n_{CO}	n_{CO_2}	n_{H_2O}	n_{H_2}	n_C	Condition
1b-a	=	=	=	>	=	>	$a_O = 1/2a_H$
1b-b	=	=	>	=	>	=	$a_O = 2a_C - 2\varepsilon$
1b-c	=	=	>	>	=	=	$a_C = 1/2a_O - 1/4a_H + 3\varepsilon$
1b-d	=	>	=	=	>	=	$a_O = a_C + \varepsilon$
1b-e	=	>	=	>	=	=	$a_O = a_C + 1/2a_H - 3\varepsilon$
1b-f	>	=	=	>	=	=	$a_H = 4a_C + 2a_O - 12\varepsilon$
1b-g	>	=	>	=	=	=	$a_C = 1/2a_O + 1/4a_H - \varepsilon$
1b-h	>	>	=	=	=	=	$a_C = a_O + 1/4a_H - 3\varepsilon$

3 | RESULTS AND DISCUSSION

The generality of the presented conceptual framework for equilibrium calculations enables the unification of equilibrium calculations in a single format that is applicable to a wide variety of industrial processes, including SMR, EESMR, DMR, and SM.

Table 2 below identifies, for each of the aforementioned industrial processes, typical operating condition ranges, and the boundaries of the attainable regions (AR) in $a_H - a_O$ space that can be generated for the typical ranges of feed compositions. The AR's of each of the processes are subsequently plotted in $a_H - a_O$ space, as shown in Figure 2.

As can be seen from Table 3 and Figure 2, the AR's for SMR and DMR are linear segments, while those for EESMR and SM are triangles. In fact, the SMR AR is simply one of the sides of the EESMR AR, as SMR is simply the limiting case of EESMR with no CO in the feed, and thus no energetic enhancement. Significant (albeit different) portions of the DMR AR are subsets of the EESMR and SM AR's. Consider for example DMR being carried out with feed ratio of CO_2/CH_4 = 5/3. This corresponds to $(a_H, a_O) = (2/5, 1/3)$, which in turn corresponds to EESMR with feed ratios of CO/CH_4 = 5, H_2O/CH_4 = 5/2 and SM with feed ratios of CO/H_2 = 1, CO_2/H_2 = 1/3.

Having established the unifying characteristics of the presented conceptual framework, Theorem 2 is subsequently used to identify the global optimum of the Gibbs free energy minimization problem for $\varepsilon = 10^{-9}$, and for temperature-pressure combinations from the lists T (K) $\in$ {550, 800, 1100, 1400}, P(MPa/0.1 MPa) $\in$ {1, 10, 30}, for which the gaseous species' compressibility factors are found to be between 0.93 and 1.02 using data from the NIST Thermophysical Properties of Fluid Systems Database,[21] validating the employed ideal gas assumption. The obtained results are presented in Figures 3-5 depicting iso $- a_O$, SMR, EESMR, DMR, and SM lines of the mole over total atom-mol normalized ratios of the ith species n_i, $i \in$ {CO, CO_2, CH_4, H_2, H_2O, C} as functions of a_H, for a a_H discretization of 0.01.

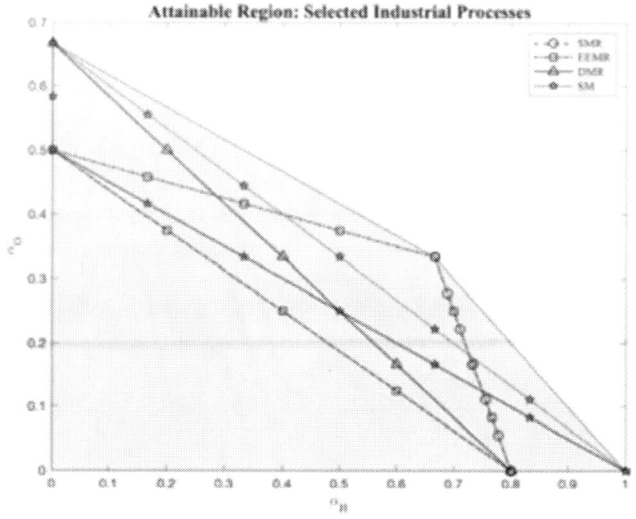

FIGURE 2 Attainable regions in $a_H - a_O$ space for four industrial processes

TABLE 2 Feasible, regular cases with at most three active constraints; (=) signifies $n_i = \varepsilon$

Case	n_{CH_4}	n_{CO}	n_{CO_2}	n_{H_2O}	n_{H_2}	n_C	Order	Case	n_{CH_4}	n_{CO}	n_{CO_2}	n_{H_2O}	n_{H_2}	n_C	Order
1	>	>	>	>	>	>	4th	21	>	>	>	=	>	=	4th
2	=	>	>	>	>	>	3rd	22	>	>	>	>	=	=	4th
3	>	=	>	>	>	>	4th	23	=	=	=	>	>	>	Explicit
4	>	>	=	>	>	>	3rd	24	=	=	>	=	>	>	Explicit
5	>	>	>	=	>	>	4th	25	=	>	=	=	>	>	Explicit
6	>	>	>	>	=	>	4th	26	=	=	>	>	=	>	Explicit
7	>	>	>	>	>	=	8th	27	=	>	=	>	=	>	Explicit
8	=	=	>	>	>	>	3rd	28	>	=	=	>	=	>	Explicit
9	=	>	=	>	>	>	4th	29	>	=	>	=	=	>	Explicit
10	>	=	=	>	>	>	4th	30	>	>	=	=	=	>	Explicit
11	=	>	>	=	>	>	2nd	31	=	=	>	>	>	=	Explicit
12	>	=	>	=	>	>	2nd	32	=	>	=	>	>	=	Explicit
13	>	>	=	=	>	>	2nd	33	>	=	=	>	>	=	Explicit
14	=	>	>	>	=	>	2nd	34	=	>	>	=	>	=	Explicit
15	>	=	>	>	=	>	2nd	35	>	=	>	=	>	=	Explicit
16	>	>	=	>	=	>	3rd	36	>	>	=	=	>	=	Explicit
17	>	>	>	=	=	>	2nd	37	=	>	>	>	=	=	Explicit
18	=	>	>	>	>	=	3rd	38	>	=	>	>	=	=	Explicit
19	>	=	>	>	>	=	7th	39	>	>	=	>	=	=	Explicit
20	>	>	=	>	>	=	4th	40	>	>	>	=	=	=	Explicit

TABLE 3 Process feed compositions, operating conditions, and attainable regions

Process	Feed composition	T (K)[a]	P (MPa)[a]	$a_H - a_O$ attainable region boundary
SMR	$\{CH_4, XH_2O\}$ $X \in (0,\infty)$	773–1273	0.1–4	$(a_H, a_O) = ((4+2X)/(5+3X), X/(5+3X)) \in \{(a_H, (4-5a_H)/2);\ a_H \in [2/3, 4/5]\}$
EESMR	$\begin{Bmatrix} CH_4, \\ XCO, \\ YH_2O \end{Bmatrix}$ $X \in (0,\infty)$, $Y \in (0,\infty)$	773–1273	0.1–4	$\begin{pmatrix} a_H, \\ a_O \end{pmatrix} = \begin{pmatrix} \frac{4+2Y}{5+2X+3Y}, \\ \frac{X+Y}{5+2X+3Y} \end{pmatrix} \in \begin{Bmatrix} \left\{\left(a_H, \frac{4-5a_H}{2}\right);\ a_H \in \left[\frac{2}{3}, \frac{4}{5}\right]\right\} \\ \cup \left\{\left(a_H, \frac{2-a_H}{4}\right);\ a_H \in \left[0, \frac{2}{3}\right]\right\} \\ \cup \left\{\left(a_H, \frac{4-5a_H}{8}\right);\ a_H \in \left[0, \frac{4}{5}\right]\right\} \end{Bmatrix}$
DMR	$\begin{Bmatrix} CH_4, \\ XCO_2 \end{Bmatrix}$ $X \in (0,\infty)$	723–1373	0.1–1	$(a_H, a_O) = \left(\frac{4}{5+3X}, \frac{2X}{5+3X}\right) \in \left\{\left(a_H, \frac{4-5a_H}{6}\right);\ a_H \in \left[0, \frac{4}{5}\right]\right\}$
SM	$\begin{Bmatrix} H_2, \\ XCO, \\ YCO_2 \end{Bmatrix}$ $X \in (0,\infty)$, $Y \in (0,\infty)$	473–873	0.1–1.5	$\begin{pmatrix} a_H, \\ a_O \end{pmatrix} = \begin{pmatrix} \frac{2}{2+2X+3Y}, \\ \frac{X+2Y}{2+2X+3Y} \end{pmatrix} \in \begin{Bmatrix} \left\{(0, a_O);\ a_O \in \left[\frac{1}{2}, \frac{2}{3}\right]\right\} \cup \\ \left\{\left(a_H, \frac{2-2a_H}{3}\right);\ a_H \in [0,1]\right\} \\ \cup \left\{\left(a_H, \frac{1-a_H}{2}\right);\ a_H \in [0,1]\right\} \end{Bmatrix}$

[a]Obtained from Reference 12.

The importance of irregular points is then discussed, and points with ε-active constraints at the optimum are presented in Figure 6. Finally, the carbon formation boundary in (a_H, a_O) space is also computed, and the obtained results are presented in Figure 7.

Figure 3 presents the mole over total atom-mol normalized ratios of the ith species at T(K) = 550, P(MPa/0.1 MPa) = 1. Close examination of Figure 3 suggests that in the range $8\varepsilon < a_H < 0.5$, and for all $a_O \in \{0.1, 0.3, 0.5\}$, CO, and H_2 are negligible, while in the range $0.75 < a_H < 0.9$ and for $a_O = 0.1$, H_2 is an increasing function of a_H. CH_4 exhibits the maxima $n_{CH_4} \in \{0.14, 0.05, 0.01\}$ at $a_H \in \{0.75, 0.58, 0.30\}$ for $a_O \in \{0.5, 0.3, 0.1\}$. H_2O on the other hand is an increasing function of a_H, while CO_2, C are decreasing functions of a_H, whose values are higher (lower) for increasing values of a_O.

Focusing on the SMR process curves, it is clear that this temperature and pressure is a poor SMR choice, as there is little H_2 present. The equilibrium mixture for SMR under these operating conditions contains mostly H_2O and CH_4. It is important to note that the SMR curves shown in Figure 3 are identified as the Gibbs free energy minima by different Table 1 Cases in different a_H ranges. Indeed, Case 19 is identified as the SMR global optimum of ν_ε from $a_H \cong 0.67$ to $a_H \cong 0.69$, in which $n_C = 0$ and $n_{CO} = \varepsilon$. Between $a_H \cong 0.69$ and $a_H \cong 0.77$, Case 7 yields the optimum in which CO is formed in greater than infinitesimal amounts. As a_H is further increased, Case 1 is identified as the global optimum in which all species are present in larger than infinitesimal amounts.

Similarly, the three EESMR process boundary curves suggest that EESMR should not be conducted at this temperature and pressure. One of the boundaries corresponds to SMR and thus the above discussed results also hold. Along the CO—H_2O EESMR boundary, as a_H increases, C decreases (and falls to zero at $a_H \cong 0.50$), while H_2O and H_2 increase (though the amount of H_2 is two orders of magnitude smaller than that of H_2O). CO_2, C, and H_2O are the dominant products along most of this boundary. Case 1 yields the global optimum of ν_ε when C is formed, while Case 7 is identified as the optimum in the absence of carbon formation. Along the CO - CH_4 EESMR boundary, carbon is always produced, reaching a maximum of $n_C \cong 0.28$ at $a_H \cong 0.25$, while all optima along this boundary are identified by Case 1.

The DMR process curve also suggests that this temperature and pressure would be a poor DMR choice, as C, H_2O rather than H_2, CO formation is favored. The amounts of C, H_2O formed reach maxima at $a_H \cong 0.5$, while CH_4, H_2 (CO_2, CO) increase (decrease) monotonically with a_H. Case 1 is identified as the optimum of ν_ε throughout the DMR attainable region.

The two nontrivial SM process boundary curves ($a_H \neq 0$) suggest that SM operating with hydrogen-lean feeds produces significant amounts of CO_2, C. As a_H increases, hydrogen atoms partition competitively between CH_4, H_2O, with H_2O being the more dominant product. A maximum of $n_{CH_4} \cong 0.12$, with a similar amount of H_2O is achieved on the CO—H_2 SM boundary for $a_H \cong 0.72$. Under

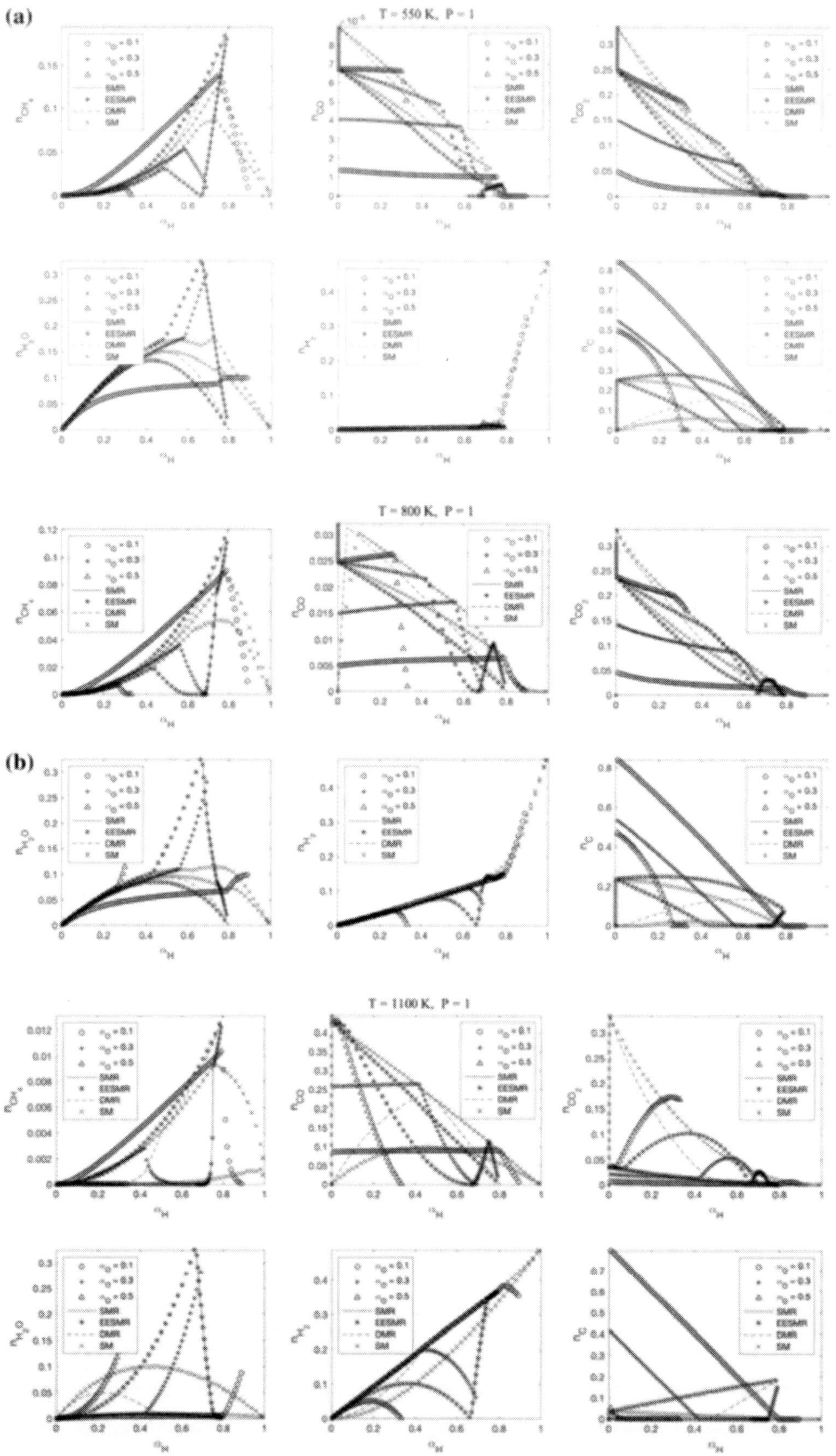

FIGURE 3 (a) Normalized molar species ratios, n_i, as a function of a_H for T = 550 K, p = 1. (b) Normalized molar species ratios, n_i, as a function of a_H for T = 800 K, p = 1. (c) Normalized molar species ratios, n_i, as a function of a_H for T = 1100 K, p = 1. (d) Normalized molar species ratios, n_i, as a function of a_H for T = 1400 K, p = 1

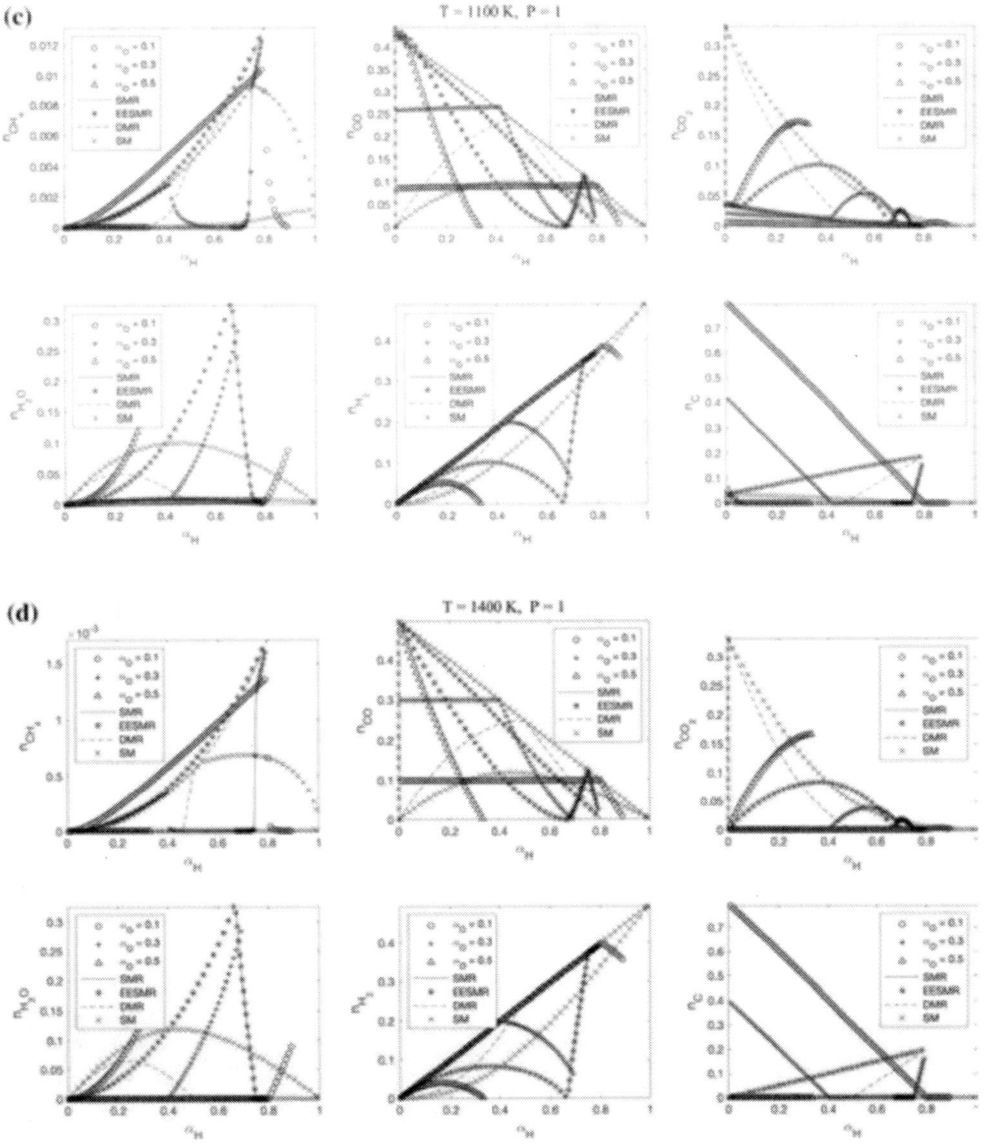

FIGURE 3 (Continued)

hydrogen-lean feeds along this boundary (small a_H), Case 1 yields the optimum of ν_r. As hydrogen is added to the feed, Case 7 yields the optimum from $a_H \cong 0.74-0.84$, Case 3 from $a_H \cong 0.84-0.90$, and Case 10 from $a_H \cong 0.90-0.99$. As for the CO_2—H_2 SM boundary, it is seen that the dominant equilibrium components are CO_2, C, similar to the CO—H_2 SM boundary. The maximum CH_4 amount produced along this boundary is lower, yielding $n_{CH_4} \cong 0.09$ at $a_H \cong 0.72$. Case 1 also yields the optimum as H_2 is added to a feed of CO_2 (CO) until $a_H \cong 0.66$. As a_H increases further, Case 7, Case 3, and Case 10 yield the optimum.

As depicted on the SMR plot of n_{H_2} in Figure 3b, conversion of CH_4 and H_2O into H_2 and carbon oxides at $T = 800$ K, reaches a maximum of $n_{H_2} \cong 0.144$ at $a_H \cong 0.72$. H_2O and H_2 are the dominant equilibrium products under these conditions, and coking is not an issue for $a_H < 0.73$. Case 7 (Case 1) is identified as the optimum for feedstocks with a_H less than (greater than) 0.73. Thus, elevated temperatures are favorable for H_2 production in SMR since the equilibrium conversion of CH_4 is much greater at $T = 800$ K than $T = 550$ K at $P = 1$. Hydrogen is also a dominant equilibrium product under EESMR, though carbon formation occurs for a wide range of operating conditions: all feedstocks along the CO—CH_4 boundary form carbon, and feedstocks with $a_H < 0.44$ form carbon along the CO—H_2O boundary (through Case 1). DMR under these conditions still yields formation of solid carbon throughout the attainable region via Case 1, as seen on the plot of n_C, so higher temperatures should be pursued for this process. The maximum obtainable amounts of methane under SM fall as temperature is increased according to the plot of n_{CH_4}, and thus lower temperatures should be chosen to operate this process.

At the even higher temperature of $T = 1100$ K, the SMR plot of n_{H_2} in Figure 3c indicates that the coking-free H_2 production increases

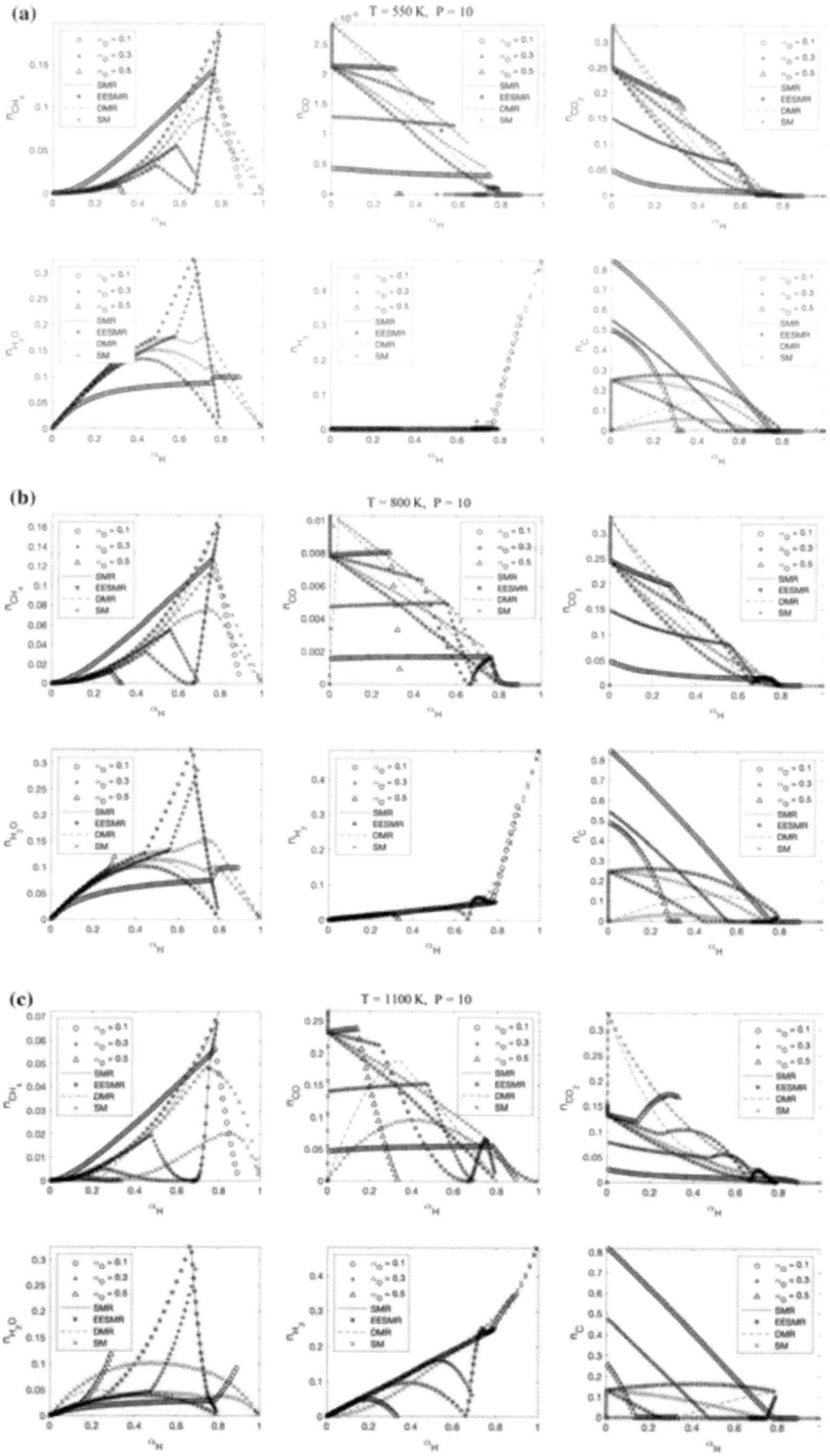

FIGURE 4 (a) Normalized molar species ratios, n_i, as a function of a_H for T = 550 K, p = 10. (b) Normalized molar species ratios, n_i, as a function of a_H for T = 800 K, p = 10. (c) Normalized molar species ratios, n_i, as a function of a_H for T = 1100 K, p = 10. (d) Normalized molar species ratios, n_i, as a function of a_H for T = 1400 K, p = 10

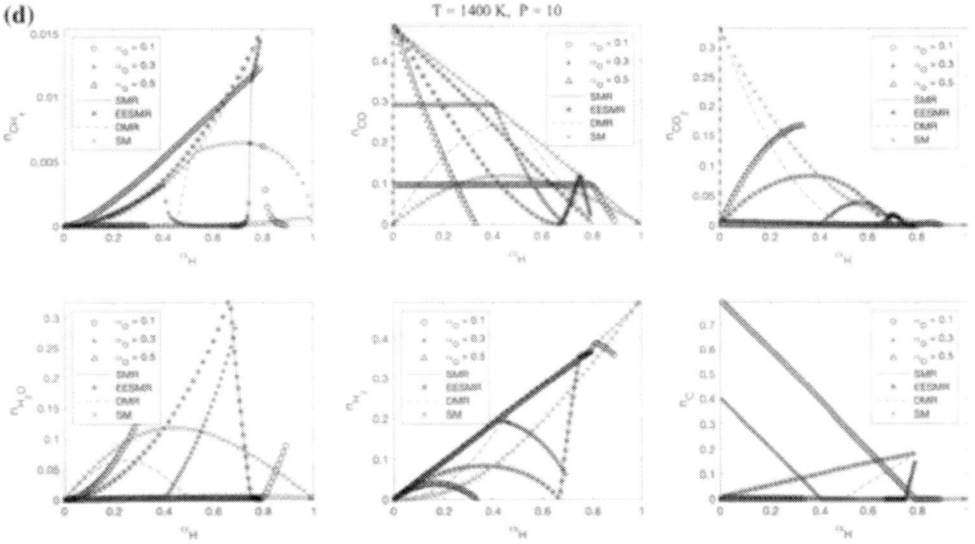

FIGURE 4 (Continued)

to $n_{H_2} \cong 0.35$ at $a_H \cong 0.74$, yielding a product stream comprised of mainly H_2 and CO in a 3.5:1 ratio. Similarly, EESMR at these conditions produces reasonable yields of H_2 without carbon formation, though coking is still an issue for feeds lean in H_2O, as seen in the plot of n_C. The DMR plot shows a maximum of $n_{H_2} \cong 0.21$ at $a_H \cong 0.46$ with no solid carbon formed, though higher H_2 production is possible with coking. The equilibrium product under these conditions is a near equimolar stream of H_2 and CO_2. SM produces meager yields of CH_4 under these elevated temperatures, though carbon formation is less prevalent.

As shown in the n_{H_2} SMR plot in Figure 3d, H_2 production as T = 1400 K is similar to that at T = 1100 K, with a carbon-free maximum at $a_H \cong 0.74$. A substantial amount of H_2 can be produced in the EESMR process without carbon formation, with coking occurring for low amounts of H_2O in the feed. DMR achieves a maximum coking-free H_2 production of $n_{H_2} \cong 0.23$ at $a_H \cong 0.47$. Coking occurs in DMR for larger a_H, as shown in the plot of n_C. Once again, the n_{CH_4} plot shows that paltry amounts of CH_4 are formed in SM under this (T, P) combination.

Summarizing Figure 3a–d, it is shown that high temperatures are good SMR operating conditions, since H_2 production without coking is maximized at (T, P) = (1 400 K, 1). Similarly, EESMR produces a substantial amount of H_2 at elevated temperatures, while DMR attains a maximum yield of coking-free H_2 at 1400 K. On the other hand, at p = 1, SM should be operated at low temperatures (i.e., T = 550 K) to obtain meaningful amounts of CH_4.

Similar analysis can be conducted vis-à-vis the trends of equilibrium species production and carbon formation as temperature increases at higher pressures. Plots are provided below and analysis is conducted subsequently.

Equilibrium compositions at P = 10, as depicted in Figure 4a–d, reveal again that H_2 production through SMR and DMR is maximized at higher temperatures, as shown in the plots of n_{H_2}. The maximum coking-free H_2 production for both SMR and DMR at P = 10 occurs at the highest temperature in the range considered, T = 1,400 K. For SMR, the maximum coking-free H_2 production at (T, P) = (1,400 K, 10) is less than the maximum yielded at (T, P) = (1400 K, 1), suggesting that lower pressures would be better choices of operating conditions. Similarly, hydrogen production in DMR slightly decreases as pressure is increased. Production of H_2 without coking in EESMR at P = 10 is maximized at T = 1100 K. Slight increases in CH_4 production in SM are observed as the pressure is increased, as seen in the plots of n_{CH_4} in Figure 4a–d, with a maximum occurring at T = 550 K for P = 10.

Figure 5a–d depict the equilibrium compositions at the highest pressure considered, P = 30. The production of H_2 for both SMR and DMR is maximized at T = 1400 K, though the production levels are still lower than the optima at lower pressures. Maximal SM production of CH_4 remains fairly flat as P is increased to 30. H_2 production without carbon formation for DMR decreases for all temperatures as P is raised to 30.

Figures 3–5 contain a wealth of insight into the equilibrium concentrations of each of the six species spanning a wide range of temperatures, pressures, and feed compositions used in industrial processes. CH_4 is most often present at equilibrium for lower temperatures, whereas increases in temperature cause the hydrogen to exist in its gaseous diatomic state and the carbon to be incorporated into CO_2. It is evident that temperature has a pronounced effect on the equilibrium formation of CO: it is a minor equilibrium component at low temperatures across the entire pressure range considered, whereas it becomes a dominant species in the gas phase at higher temperatures. The decomposition of CO into CO_2 and carbon is prevalent at low temperatures, and thus care should be taken to ensure that coking does not occur, especially in SM. H_2 production in SMR and EESMR is optimized at high temperatures and low to intermediate pressures.

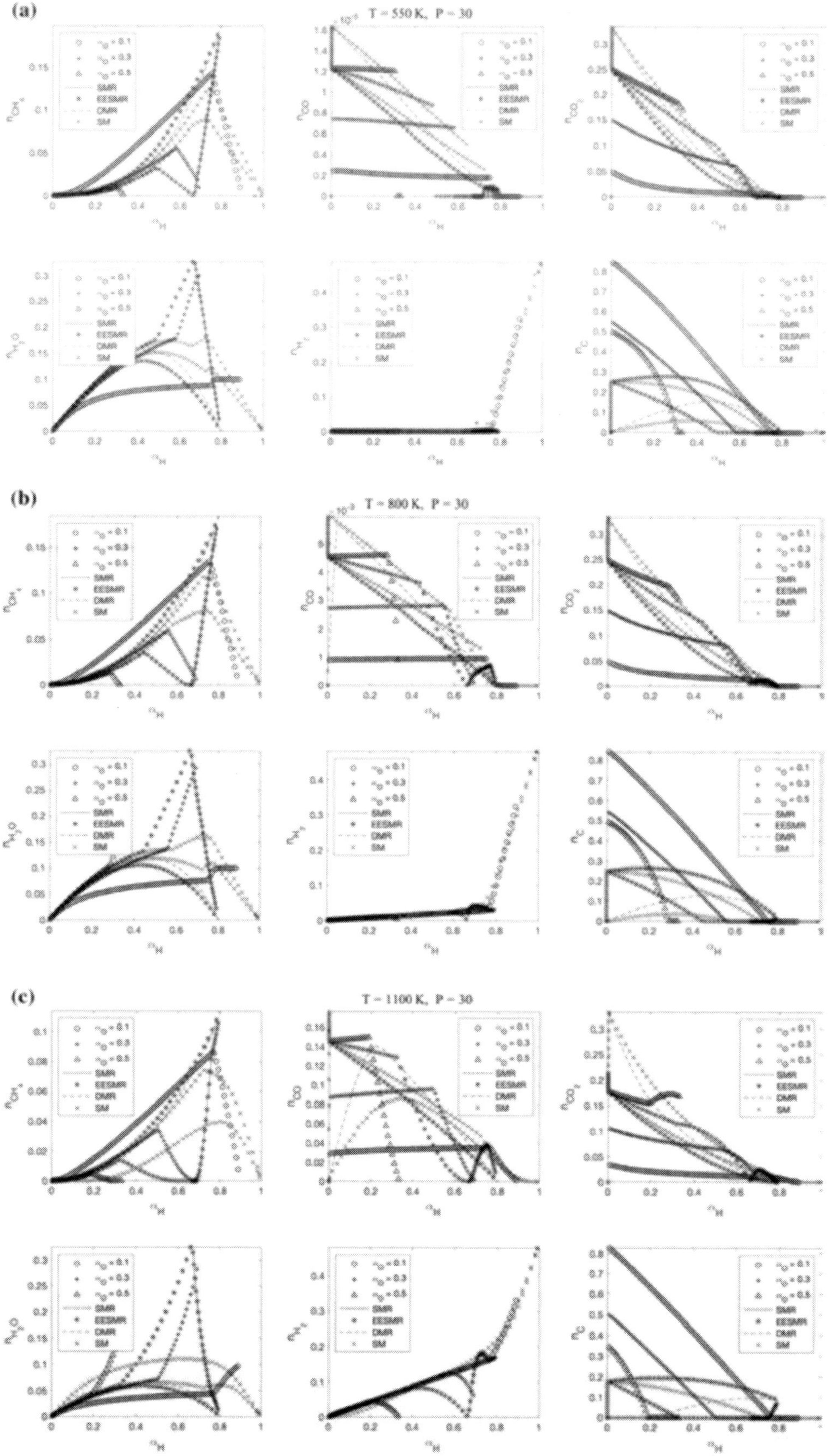

FIGURE 5 (a) Normalized molar species ratios, n_i, as a function of a_H for T = 550 K, p = 30. (b) Normalized molar species ratios, n_i, as a function of a_H for T = 800 K, p = 30. (c) Normalized molar species ratios, n_i, as a function of a_H for T = 1100 K, p = 30. (d) Normalized molar species ratios, n_i, as a function of a_H for T = 1400 K, p = 30

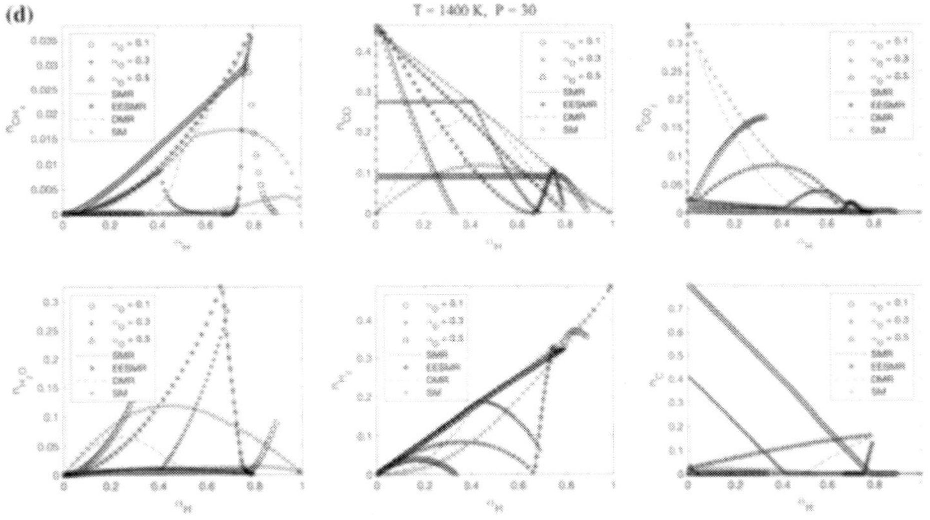

FIGURE 5 (Continued)

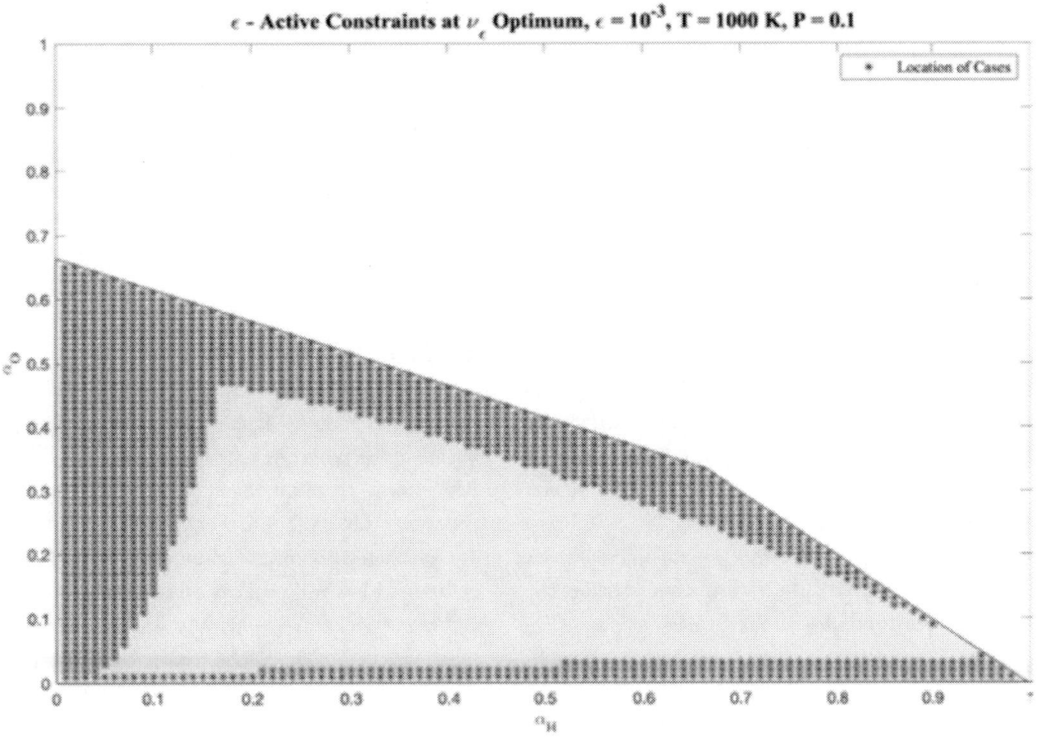

FIGURE 6 $a_H - a_O$ locations with ε-active constraints at ν_ε (T = 1000 K, p = 1, $\varepsilon = 10^{-3}$)

Having presented the equilibrium mole over total atom-mol normalized ratios for all species, at all considered temperatures, and pressures, next the importance of the choice of ε is discussed. In the majority of the above presented results obtained for a value of $\varepsilon = 10^{-9}$, the global minimum ν_ε occurs in one of two cases: when carbon is formed, Case 1 ($n_i > \varepsilon \; \forall \, i$ = {CH_4, CO, CO_2, H_2O, H_2}, $n_C > 0$); and when carbon is not formed, Case 7 ($n_i > \varepsilon \; \forall \, i$ = {CH_4, CO, CO_2, H_2O, H_2}, $n_C = 0$). However, use of a larger value of ε leads to a bounty of cases other than Case 1, or Case 7 yielding the global optimum, especially around the boundaries of the feasible region. Indeed, for $\varepsilon = 10^{-3}$, T = 1000 K, and P = 1, Figure 6 illustrates all points in $a_H - a_O$ space for which ε-active cases determine the global minimum of ν_ε. As an example, Case 4 yields the global minimum for a wide range of feed compositions as $a_O \to 0$. This is to be expected, since under these oxygen atom lean conditions, there are only enough oxygen atoms in the mixture to form CO, but not CO_2, above the ε level considered.

It is also worth noting that irregular points near the corners of the region, as identified in Theorem 1b, can yield the global minimum. For

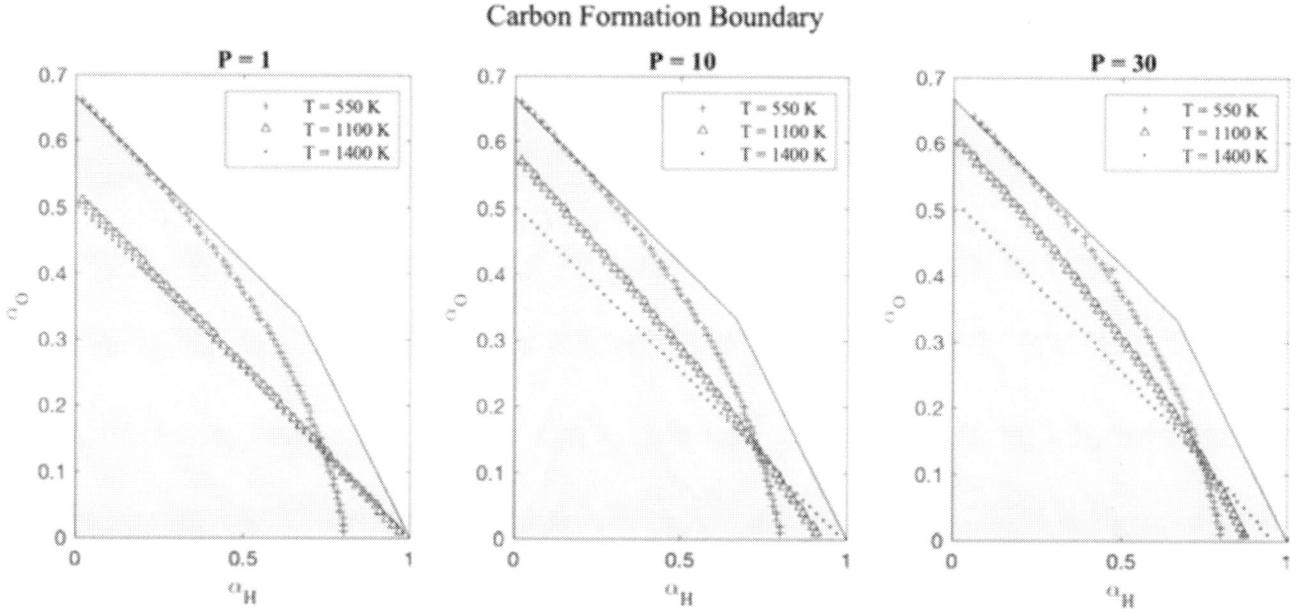

FIGURE 7 Carbon formation boundaries and regions (below boundaries)

example, at $(T, P) = (550\ \text{K},\ 1)$ with $\varepsilon = 10^{-9}$, consider the irregular point in Theorem 1a-a corresponding to $(a_H, a_O) = (9\varepsilon, 9\varepsilon/2)$. The optimum identified through the 40 feasible cases, is obtained by Case 29, which yields $\nu_\varepsilon = -986.244$, whereas the irregular case 1a-a yields $\nu_\varepsilon = -986.248$ and is therefore the global optimum, signifying that only H_2O and C are present at equilibrium in meaningful amounts. Similar assessments can be made for other irregular cases near the boundaries of the feasible region.

The carbon formation boundary is plotted in Figure 7 as a function of a_H and a_O, for $T = 550$ K, $T = 1100$ K, and $T = 1400$ K, and for each of the pressures considered. Carbon formation boundaries for intermediate temperatures manifest themselves as curves in between the two extreme curves obtained for $T = 550$ K and $T = 1400$ K.

It is evident from Figure 7 that the carbon formation boundaries at the lowest temperatures are markedly nonlinear, approaching the axes corresponding to feedstocks of CH_4 along the a_H axis and CO_2 along the a_O axis. As T increases at a given P, the carbon boundary straightens and approaches a limiting line intersecting the axes at $(a_H, 0) = (1, 0)$ and $(0, a_O) = (0, 0.5)$. We also note that the boundaries appear to converge near $(a_H, a_O) = (0.74, 0.14)$, though the three curves do not intersect at a single point.

It is important to note that many industrial reactions do not achieve equilibrium due to a variety of reasons, such as kinetic limitations. One possible avenue to extending the applicability of equilibrium calculations to circumstances where equilibrium is not industrially attainable, is the "approach temperature to equilibrium" concept,[22] whereby equilibrium calculations at a given temperature provide reasonable approximations to industrially obtainable results at a different (typically higher) temperature. In that context, Figure 7, demonstrates the effect of temperature on coke formation at equilibrium conditions. The carbon formation region is shown to be larger (smaller) for small (large) values of a_H at lower temperatures. In turn, this suggests that at industrial conditions, one should expect higher (lower) levels of carbon formation, for small (large) values of a_H, than the equilibrium calculations at the same temperature predict.

4 | CONCLUSIONS

A general conceptual framework, based on Gibbs free energy minimization, has been presented that can be used to carry out in a unified manner chemical-phase equilibrium computations for systems containing CH_4, CO, CO_2, H_2O, H_2, and potentially solid carbon. The presented approach can be used to carry out equilibrium computations for a wide variety of processes, including SMR, EESMR, DMR, and SM, and can identify the attainable regions for each of these processes. The Gibbs free energy based minimization problem was formulated in atom ratio space, and a Theorem was proved that explicitly identified the finite and bounded region in atom ratio space for which the aforementioned minimization problem is feasible. Additional Theorems identified conditions that must be satisfied by regular and irregular points that are candidate optima. An algorithm was then presented that is guaranteed to identify the global minimum of a feasible minimization problem, for a wide variety of feedstock atom ratios, temperatures and pressures, yielding the total atom-mol normalized molar ratios for all species. The algorithm is based on the identification of all roots of a number of single variable, polynomial equations, and the comparison of the resulting Gibbs free energy values corresponding to each of these roots. The algorithm's importance was illustrated near the boundaries of the feasible region. It was shown that for the set of temperatures and pressures considered, operation without carbon formation should be carried out at high temperatures and low pressures for steam methane reforming, energetically-enhanced methane reforming, and dry methane reforming, and that syngas methanation should occur at low temperatures and intermediate pressures. In particular, the carbon formation boundary in atom ratio space was identified, for a range of temperatures and pressures. It is also important to note that many industrial processes operate under non-equilibrated conditions for a variety of reasons, including practical limitations. Under such regimes, the Gibbs minimization method as outlined in the present work does not hold since it identifies only the equilibrium compositions and ignores any transient effects on product compositions and flow rates. Future research efforts will focus on extending the proposed methodology to the equilibrium analysis of mixtures involving different species.

NOTATION

$\bar{a}$ (total atom-mol)	total number of C, H, O atom-mols present in system
$\bar{a}_j$ (total atom j-mol)	number of atom-mols of element j present in system
a_C (C atom - mol/total atom - mol)	fraction of carbon atom-mol over total atom-mol present in system
a_H (H atom - mol/total atom - mol)	fraction of hydrogen atom-mol over total atom-mol present in system
a_O (O atom - mol/total atom - mol)	fraction of oxygen atom-mol over total atom-mol present in system
f	optimization problem objective function
h	optimization problem equality constraint function
g	optimization problem inequality constraint function
$G\ (T,P,\{\bar{n}_i\}_{i\in S})\ (J)$	total Gibbs free energy of the system
$G_i^{\circ}(T)$ (J/mol i)	standard molar Gibbs free energy of the ith species at temperature T
$\Delta H_{f,i}^{\circ}$ (J/mol i)	standard molar enthalpy of formation of the ith species at 298.15 K, 1 bar
$\Delta H_{vap,i}^{\circ}$ (J/mol i)	standard molar enthalpy of vaporization of the ith species at its standard boiling point, 1 bar
M_i	Lagrange multiplier exponents associated with the optimization formulation
n_i(mol i/total atom - mol)	ratio of species i mol over total atom-mol present in system
$\bar{n}_i$ (mol$i$)	total number of mols of species i present in system
P (MPa/0.1 MPa)	dimensionless operating pressure normalized over standard pressure of 1 bar
R(J/mol - K) = 8.3145	universal gas constant
S	set of species $S \triangleq$ {CO, CO2, CH4, H2, H2O, C}
$\Delta S_{f,i}^{\circ}$ (J/(mol K))	standard molar entropy of formation of ith species 298.15 K, 1 bar
$\Delta S_{vap,i}^{\circ}$ (J/(mol K))	standard molar entropy of vaporization of the ith species at its standard boiling point, 1 bar
T (K)	process temperature
$T_{BP,i}$ (K)	standard boiling point of ith species, at 1 bar
X, Y	constants parameterizing process feed composition

GREEK LETTERS

$\alpha, \beta, \gamma, \mathrm{A}, \mathrm{B}, \Gamma$	equilibrium constant-related functions of temperature
ε	infinitesimally small, positive constant
λ	Lagrange multipliers associated with equality constraints
μ	Lagrange multipliers associated with inequality constraints
ν	optimization problem's optimum objective function value
ν_{*}(J/atom - mol)	total atom-mol normalized Gibbs free energy optimum value
$\bar{\nu}$ (J)	total Gibbs free energy optimum value
$\nu_{j,i}$	number of atoms of element j in species i

ORCID

Demetrios Chaconas https://orcid.org/0000-0003-2552-0478
Vasilios I. Manousiouthakis https://orcid.org/0000-0003-1548-0097

REFERENCES

1. Gibbs JW. A method of geometrical representation of the thermodynamic properties of substances by means of surfaces. *Trans Connecticut Acad Arts Sci*. 1873;1:382-404.
2. Bartholomew CH. Carbon deposition in steam reforming and methanation. *Catal Rev*. 1982;24:67-112.
3. Udengaard NR. Hydrogen production by steam reforming of hydrocarbons. *ACS Div Fuel Chem Preprints*. 2004;49:906-907.
4. Annesini MC, Piemonte V, Turchetti L. Carbon formation in the steam reforming process: a thermodynamic analysis based on the elemental composition. *Chem Eng Trans*. 2007;11:21-26.
5. Manousiouthakis VI, Albassam AM, Conner JA. Energetically enhanced reforming process. US 2017/0001862 A1; 2017.
6. Pichardo P, Manousiouthakis VI. Infinite DimEnsionAl state-space as a systematic process intensification tool: energetic intensification of hydrogen production. *Chem Eng Res Des*. 2017;120:372-395.
7. Pichardo P, Manousiouthakis VI. On the intensification of natural gas-based hydrogen production utilizing hybrid energy resources. *Smart Sustain Manuf Syst*. 2018;2(2):1-24.
8. Pichardo PA, Manousiouthakis VI. Intensified energetically enhanced steam methane reforming through the use of membrane reactors. *AIChE J*. 2020;66:e16827.

9. Al-Bassam AM, Conner JA, Manousiouthakis VI. Natural-gas-derived hydrogen in the presence of carbon fuel taxes and concentrated solar power. *ACS Sustain Chem Eng*. 2018;6:3029-3038.
10. Nematollahi B, Rezaei M, Lay EN, Khajenoori M. Thermodynamic analysis of combined reforming process using Gibbs energy minimization method: in view of solid carbon formation. *J Nat Gas Chem*. 2012;21:694-702.
11. Cao P, Adegbite S, Wu T. Thermodynamic equilibrium analysis of CO_2 reforming of methane: elimination of carbon deposition and adjustment of H2/CO ratio. *Energy Procedia*. 2017;105:1864-1869.
12. Hiller H, Reimert R, Marschner F, et al. *Gas Production*. Weinheim, Germany: Ullmann's Encyclopedia of Industrial Chemistry; 2006.
13. Tavares MT, Alstrup I, Bernardo CA, Rostrup-Nielsen JR. Carbon formation and CO methanation on silica-supported nickel and nickel–copper catalysts in CO + H_2 mixtures. *J Catal*. 1996;158:402-410.
14. Sang JC, Hae JK, Kim SJ, Park SB. Adsorbed carbon formation and carbon hydrogenation for CO_2 methanation on the Ni(111) surface: ASED-MO study. *Bull Kor Chem Soc*. 2005;26:1682-1688.
15. Gao J, Wang Y, Ping Y, et al. A thermodynamic analysis of methanation reactions of carbon oxides for the production of synthetic natural gas. *RSC Adv*. 2012;6:2358-2368.
16. Li M. Equilibrium calculation of gaseous reactive systems with simultaneous species adsorption. *Ind Eng Chem Res*. 2008;47:9263-9271.
17. Li M. Thermodynamic analysis of adsorption enhanced reforming of ethanol. *Int J Hydrogen Energy*. 2009;34:9362-9372.
18. Luenberger DG. *Optimization by Vector Space Methods*. New York, NY: Wiley; 1969.
19. Luenberger DG, Ye Y. *Linear and Nonlinear Programming*. 4th ed. New York, NY: Springer International; 2016.
20. Smith JM, Van Ness HC, Abbott MM. *Introduction to Chemical Engineering Thermodynamics*. 8th ed. New York, NY: McGraw-Hill Education; 2018.
21. Burgess Jr. DR. Thermophysical Properties of Fluid Systems in NIST Chemistry WebBook, NIST Standard Reference Database Number 69. National Institute of Standards and Technology, Gaithersburg MD, 20899. Retrieved May 10, 2020. https://doi.org/10.18434/T4D303.
22. White CW III, Seider WD. Approach to chemical equilibrium. *AIChE J*. 1983;27:466-471.

SUPPORTING INFORMATION

Additional supporting information may be found online in the Supporting Information section at the end of this article.

How to cite this article: Chaconas D, Pichardo P, Manousiouthakis IV, Manousiouthakis VI. Equilibrium analysis of CH_4, CO, CO_2, H_2O, H_2, C mixtures in C–H–O atom space using Gibbs free energy global minimization. *AIChE J*. 2020; e17052. https://doi.org/10.1002/aic.17052

609444 A Nonsmooth Approach to Multicontaminant Water Integration

Abstract:

Water integration methods have been widely proposed and utilized to improve sustainability both in well-established chemical processes and in new designs. These approaches can significantly reduce environmental impact by optimally reusing water to minimize fresh and wastewater utilities. However, most of these methods can only consider a single stream property, and therefore often provide infeasible results for real systems in which water reuse is limited by the presence of multiple contaminants. Solving these problems efficiently is inherently challenging because they require nonsmooth operations, both to determine the limiting contaminant in each water-using unit and also to sort streams into intervals based on these limiting concentrations. As a result, the existing approaches to solving multicontaminant problems are primarily limited to superstructures that scale exponentially with the number of resource streams and are not well-equipped to handle large-scale problems or to be embedded in programs for simultaneous process optimization.

To improve scaling and decrease complexity, we present a new approach to solving the multicontaminant water integration problem that takes advantage of its nonsmooth structure by using compact, explicitly nonsmooth expressions. To select streams by concentration interval, we have extended our previous work on the nonsmooth integration operator, which uses a system of two nonsmooth equations to both partition streams and describe optimality conditions for pinch-constrained resource transfer [1]. Our extended formulation considers multiple contaminants by incorporating the nonsmooth concentration scalings proposed by Alva-Argaez et al. in their superstructure approach [2]. These scalings describe the interrelated mass transfer between components and use max expressions to ensure that one component is always at its limit. When included in the nonsmooth integration operator, the resulting equation system can be combined with a process model and solved using new advances in nonsmooth equation solving [3] to simulate fixed-load, multicontaminant systems.

Using example systems, we will demonstrate that this approach uniquely enables us to solve both for optimal resource targets and for process parameters, including the contaminant concentrations present in the nonsmooth scalings. In addition, unlike other approaches, the resulting formulation requires only equation-solving methods and retains a compact system of only two equations regardless of the number of water streams. As a result, we have developed a readily adaptable approach that significantly reduces problem complexity and can provide computationally tractable solutions to a wide variety of large-scale, multicontaminant integration problems.

[1] C. J. Nielsen and P. I. Barton. Ind. Eng. Chem. Res. 59(1): 253-264, 2020.

[2] A. Alva-Argaez, A. Vallaintos, and A. Kokossis. Comput. Chem. Eng. 23: 1439-1453, 1999.

[3] K. A. Khan and P. I. Barton. Optim. Methods and Softw., 30(6): 1185-1212, 2015.

602521 A Comprehensive Investigation of Dendrite Formation in Lithium Anode Batteries: A Computational Approach for Heterogenous SEI

Abstract:

Lithium-ion batteries (LIB) are ubiquitously used as energy storage devices in electronics, communication, electric vehicles, etc. Most of the LIB employ a graphite anode, which unfortunately has a low energy density of ~ 372 mAh/g[1]. There are growing appeals for Li metal being the 'Holy-Grail' anode in LIB due to its high energy density (~ 3862 mAh/g). However, it is plagued by uncontrollable and unpredictable dendrite growth resulting in poor cycling efficiency, and in extreme cases, short-circuiting of batteries. The literature is rife with studies highlighting different factors that aid/cause dendrite formation, viz., overpotential, temperature[2], number of charge-discharge cycles, etc. However, there are very few studies addressing the complex interaction between heterogenous solid-electrolyte interface (SEI) and the growing dendrite. Hence, a detailed mechanistic study of factors affecting dendrite growth and the complicated SEI-dendrite interactions, which can help in devising mitigation strategies, is highly coveted in the materials community.

The focus of this work is to address the above-mentioned knowledge gap by proposing a first-principled mesoscopic kinetic Monte Carlo (KMC) simulation in combination with DFT studies to simulate dendrite growth in LIB. The KMC employs fundamental events, viz., Li^+ desolvation, diffusion of Li^+ in the solid SEI, and Li reduction. The desolvation rate, , accounts for the effect of different electrolytes. The Li^+ diffusion rate, , changes with different SEI species, viz., Li_2O, LiF, RoLi, etc. The Li^+ reduction rate accounts for the effect of overpotential and process temperature. The microscopic events are modelled based on known and/or new DFT studies[2,3]. Furthermore, the simulations account for the spatio-temporal evolution of the heterogenous SEI due to dendrite formation. The SEI evolution is predicted based upon the ratio of Young's moduli of different SEI species.

The KMC simulation predicts accelerated dendrite formation for the case of a realistic heterogenous SEI as compared to an ideal homogenous SEI case, showcasing the effect of heterogenous SEI. To illustrate this observation, video snapshots of the simulations were generated showing the temporal evolution of Li depositions into mossy, tree-like dendrites, which are consistent with experimental results[4]. However, at elevated temperatures and low overpotential values, the Li deposition was more uniform, even for a heterogenous SEI. The effect of SEI-dendrite interaction was evident from the simulation results which showed that softer SEI species (low Young's modulus) viz., ROLi, $ROCO_2Li$, underwent more SEI displacement as compared to rigid SEI species viz., LiF, Li_2O. Furthermore, DFT calculations were performed for the fast-screening of potential electrolytes which are suitable for mitigation of dendrite growth[3].

In summary, the work provides a comprehensive computational analysis of various factors affecting dendrite growth in a realistic heterogenous SEI environment and underlines the importance of physical and chemical interaction between the SEI and dendrites. The combination of DFT and KMC highlights the value of multiscale modelling in the formulation of mitigation strategies for dendrite formation in LIB.

Literature Cited

1. *Cheng, X.-B., Zhang, R., Zhao, C.-Z. & Zhang, Q. Toward safe lithium metal anode in rechargeable batteries: a review. Chemical reviews* ***117****, 10403–10473 (2017).*
2. *Hao, F., Verma, A. & Mukherjee, P. P. Mechanistic insight into dendrite–SEI interactions for lithium metal electrodes. Journal of Materials Chemistry A* ***6****, 19664–19671 (2018).*
3. *Qin, X., Shao, M. & Balbuena, P. B. Elucidating mechanisms of Li plating on Li anodes of lithium-based batteries. Electrochimica Acta* ***284****, 485–494 (2018).*
4. *Mehdi, B. L. et al. Observation and quantification of nanoscale processes in lithium batteries by operando electrochemical (S) TEM. Nano letters* ***15****, 2168–2173 (2015).*
5. *Wood, K. N. et al. Dendrites and pits: Untangling the complex behavior of lithium metal anodes through operando video microscopy. ACS central science* ***2****, 790–801 (2016).*

608856 Performance and Durability of Advaned Mn-Based Precious-Group Metal-Free Catalyst in Membrane Electrode Assemblies

Abstract:

Platinum-group metal (PGM) catalysts have unique advantages of high stability and catalytic activity for the oxygen reduction reaction (ORR), thus widely being used in proton exchange membrane fuel cells (PEMFCs). However, the scarcity and high cost of PGMs prohibit their extensive application in PEMFCs. Recently, PGM-free catalysts have attracted increasing attention due to their abundance, low cost and improved ORR catalytic activity. Fe based FeN_4 and Co-based CoN_4 catalysts have been widely investigated due to their high activities.[1] Unfortunately, cobalt-based materials also have limited availability and are known to be toxic, and iron-based materials have durability issues. Additionally, in a real membrane electrode assembly (MEA) environment, Fe-based catalysts are notorious for their induced Fenton's reaction[2] which generates peroxide radicals, rapidly degrading perfluorosulfonic acid (PFSA) membrane/ionomer in the MEA. To address the instability issue of Fe-based catalyst, a Mn-based catalyst has been developed[3].

In this work, novel electrode and ink formulation designs have been extensively studied to improve the MEA performance using $Mn\text{-}N_4$ catalyst coordination, which includes a freeze drying approach to increase electrode porosity and ionomer to catalyst (I/C) ratio tuning to enhance proton transport. We also integrate computational fluid dynamic (CFD) modeling results to better understand the impact of porosity and I/C ratio variation via oxygen concentration, oxygen reaction rate and electrolyte phase potential across the cathode catalyst layer.

The durability of Mn-based catalysts in the MEA level has also been studied in this work. It has been found that the Mn-based MEA is more stable than Fe-based catalysts, possibly due to higher degree of carbon graphitization and mitigated Fenton reaction. Testing protocols, oxygen concentration, relative humidity and temperature were all found to impact MEA durability. The degradation mechanism of Mn-based catalysts has been studied using synchrotron X-ray absorption spectroscopy (XAS). The oxidation and agglomeration of active sites have been found to be two major mechanisms for degradation.

Acknowledgement:

The project is financially supported by the Department of Energy's Fuel Cell Technology Office under the Grant DE-EE0008075.

Reference:

1. Wu, G.; More, K. L.; Johnston, C. M.; Zelenay, P., High-Performance Electrocatalysts for Oxygen Reduction Derived from Polyaniline, Iron, and Cobalt. *Science* **2011,** *332* (6028), 443-447.
2. Du, L.; Prabhakaran, V.; Xie, X.; Park, S.; Wang, Y.; Shao, Y., Low-PGM and PGM-Free Catalysts for Proton Exchange Membrane Fuel Cells: Stability Challenges and Material Solutions. *Advanced Materials* **2020,** *n/a* (n/a), 1908232.
3. Li, J.; Chen, M.; Cullen, D. A.; Hwang, S.; Wang, M.; Li, B.; Liu, K.; Karakalos, S.; Lucero, M.; Zhang, H.; Lei, C.; Xu, H.; Sterbinsky, G. E.; Feng, Z.; Su, D.; More, K. L.; Wang, G.; Wang, Z.; Wu, G., Atomically dispersed manganese catalysts for oxygen reduction in proton-exchange membrane fuel cells. *Nature Catalysis* **2018,** *1* (12), 935-945.

608324 Achieving High PEFC Performance at Rated Power through Modifications to Electrode Microstructure

Abstract:

Improvements in polymer electrolyte fuel cell (PEFC) electrode performance have primarily focused on catalyst and ionomer development without fully considering the importance of process variables such as ink formulation or electrode fabrication. Differences in catalyst ink formulations can lead to significant differences in kinetic activity, ionic and gas transport properties.[1–3] The catalyst ink solvent (water:alcohol) ratio is an important parameter that has been shown to dramatically affect ionomer/electrocatalyst interactions at the catalyst interface and throughout the electrode microstructure.[4] The difficulty of characterizing electrodes (especially the ionomer/catalyst interface) has limited our understanding of how the catalyst ink formulation influences electrode structure and performance. In this work, we comprehensively examine the effect of water:nPA ratio across a series of PEFC cathodes comprised of state-of-the-art commercial platinum alloy catalysts supported on high surface area carbon (PtCo/HSC) with Nafion (D2020) ionomer.[5]

Using a myriad of *in situ* electrochemical and *ex situ* characterization techniques to probe electrode properties across multiple length scales, we relate performance differences (namely O_2 transport) with changes in electrode microstructure. CO displacement chronoamperometry and electrochemical impedance spectroscopy (EIS) experiments indicate that ink solvent ratio had a minimal effect on ionomer coverage unlike earlier work on Pt/Vu electrodes. Nano-computed tomography (nano-CT) and ultra-small angle X-ray scattering (USAXS) establish that the optimal ink formulation (3:7 nPA:H_2O) coincides with smaller ionomer/catalyst aggregates and improved local O_2 transport. Using *commercially available materials* (catalyst and ionomer), a state-of-the-art membrane electrode assembly with 0.03/0.08 mg_{Pt}/cm^2 loading on the anode and cathode, respectively, achieves i) a mass activity exceeding 1 A/mg_{Pt} (0.9 $V_{iR\text{-}free}$, 150 kPa, 80°C, 100% RH, H_2/O_2), ii) 320 mA/cm^2 at 0.8 V, 150 kPa, 80°C, 100% RH, H_2/Air), and iii) > 1 W/cm^2_{elec} at rated power (0.67 V, 250 kPa, 94°C, 65% RH, H_2/Air) or < 0.11 g_{Pt}/kW_{rated}, all of which meet or exceed current DOE targets.[5,6]

References

1. Orfanidi, A., Rheinländer, P. J., Schulte, N. & Gasteiger, H. A. Ink Solvent Dependence of the Ionomer Distribution in the Catalyst Layer of a PEMFC. *J. Electrochem. Soc.* (2018). doi:10.1149/2.1251814jes
2. Wang, X. X., Sokolowski, J., Liu, H. & Wu, G. Pt alloy oxygen-reduction electrocatalysts: Synthesis, structure, and property. *Chinese Journal of Catalysis* **41**, 739–755 (2020).
3. Takahashi, S., Mashio, T., Horibe, N., Akizuki, K. & Ohma, A. Analysis of the Microstructure Formation Process and Its Influence on the Performance of Polymer Electrolyte Fuel-Cell Catalyst Layers. *ChemElectroChem* **2**, 1560–1567 (2015).
4. Van Cleve, T. *et al.* Dictating Pt-Based Electrocatalyst Performance in Polymer Electrolyte Fuel Cells, from Formulation to Application. *ACS Appl. Mater. Interfaces* **11**, 46953–46964 (2019).
5. Van Cleve, T. *et al.* Tailoring Electrode Microstructure via Ink Content to Enable Improved Rated Power Performance for PtCo/HSC Based Polymer Electrolyte Fuel Cells. *J Power Sources* [*submitted*]
6. Fuel Cell Multi-Year Research, Development and Demonstration Plan. (2012).

603695 Durable and High-Power PGM-Free Cathodes for PEM Fuel Cells

Abstract:

The urgent need to address the high-cost issue of proton-exchange membrane fuel cell (PEMFC) technologies, particularly for transportation applications, drives the development of simultaneously highly active and durable platinum group metal-free (PGM-free) catalysts and electrodes. The past decade has witnessed remarkable progress in exploring PGM-free cathode catalysts for the oxygen reduction reaction (ORR) to overcome sluggish kinetics and catalyst instability in acids. Among others, scientists have identified the newly emerging atomically dispersed transition metal (M: Fe, Co, or/and Mn) and nitrogen co-doped carbon (M-N-C) catalysts as the most promising alternative to PGM catalysts. Here, we provide a comprehensive review of significant breakthroughs, remaining challenges, and perspectives regarding the M-N-C catalysts in terms of catalyst activity, stability, and membrane electrode assembly (MEA) performance. A variety of novel synthetic strategies demonstrated effectiveness in improving intrinsic activity, increasing active site density, and attaining optimal porous structures of catalysts. Rationally designing and engineering the coordination environment of single metal MN_x sites and their local structures are crucial for enhancing intrinsic activity. Increasing the site density relies on the innovative strategies of restricting the migration and agglomeration of single metal sites into metallic clusters. Relevant understandings provide the correlations among the nature of active sites, nanostructures, and catalytic activity of M-N-C catalysts at the atomic scale through a combination of experimentation and theory. Current knowledge of the transferring catalytic properties of M-N-C catalysts to MEA performance is limited. Rationally designing morphologic features of M-N-C catalysts play a vital role in boosting electrode performance through exposing more accessible active sites, realizing uniform ionomer distribution, and facilitating mass/proton transports. We outline future research directions concerning the comprehensive evaluation of M-N-C catalysts in MEAs. The most considerable challenge of current M-N-C catalysts is the unsatisfied stability and rapid performance degradation in MEAs. Therefore, we further discuss practical methods and strategies to mitigate catalyst and electrode degradation, which is fundamentally essential to make M-N-C catalysts viable in PEMFC technologies.

596111 Microelectrode Study of Oxygen Mass Transport and Its Influence on the Performance of High Temperature Proton Exchange Membrane Fuel Cells

Abstract:

Oxygen mass transport in the cathode catalyst layer (CCL) of high temperature proton exchange membrane fuel cells (HT-PEMFCs) plays an important role in promoting the fuel cell performance. Binders, such as polybenzimidazole (PBI), polytetrafluoroethylene (PTFE) and poly(vinyl pyrrolidone) (PVP) are used to adjust O_2 mass transport in the catalyst layer (CL) of HT-PEMFCs. Up to now, many researches have reported the influence of binders loading on the performance of HT-PEMFCs. However, all of these researches are qualitative, and there are no direct and quantitative evidences to illustrate the effect of binder compositions on O_2 mass transport in CCL. In addition, phosphoric acid (PA) leaching is another crucial issue for HT-PEMFCs. It is generally agreed that PA leaching leads to the proton conductivity decline of PEM, which leads to fuel cell performance degradation. But the quantitative effect of PA leaching on O_2 mass transport in CCL has rarely been investigated.

In this work, we employ a microelectrode technique to quantify the effect of the binder composition and PA leaching on O_2 mass transport in CCL of HT-PEMFCs. A microelectrode electrochemical cell is designed and fabricated for quantitatively characterizing the O_2 mass transport coefficients in CCL of HT-PEMFCs. A series of electrochemical measurements, including high-resolution cyclic voltammetry, potential-step chronoamperometry and sampled chronoamperometry, at the nA scale is conducted on a poly(ethersulfone)-poly(vinyl pyrrolidone) (PES-PVP) binder under various conditions.

A systematic data analysis is used to successfully obtain the diffusion coefficients (D_{O2}), solubility (C_{O2}) and permeability ($D_{O2}*C_{O2}$) of O_2 in PES-PVP binders with different PVP contents. We find that the permeability ($D_{O2}*C_{O2}$) decreases as the PVP content increases, indicating that the binder composition has a considerable effect on O_2 mass transport in the binder.

We also find that the permeability first increases as the PA doping level in the binder increases from 2.1 to 5.1, but then decreases upon further increase of the PA doping level. This result indicates an optimal PA doping level in the binder.

The research methods and conclusions presented in this study pave the way for optimizing the cathode catalyst layer (CCL) structure of HT-PEMFCs, thereby realizing highly efficient O_2 mass transport in the CCL and improving the cell performance.

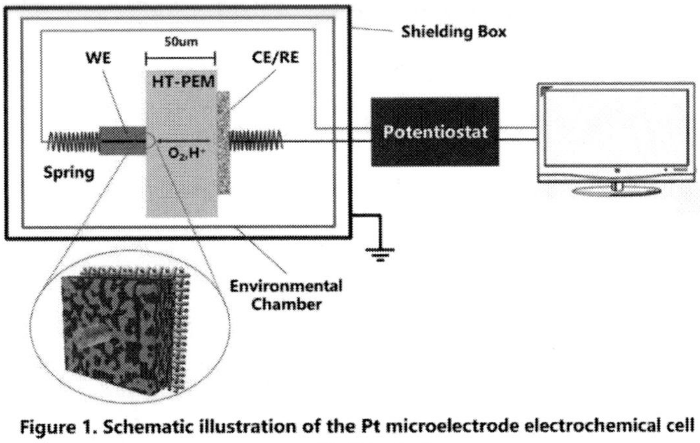

Figure 1. Schematic illustration of the Pt microelectrode electrochemical cell used in this study.

601833 Impact of Ionomer Content on Fuel-Cell Catalyst Layer Structure

Abstract:

Catalyst layers (CLs) in polymer-electrolyte fuel-cells are the micron-thick porous structures fabricated using inks consisting of carbon particles with Pt nanoparticles and ionomer dispersed in a solvent. The various phases are essential to transport reactant gases, protons and electrons to/from the Pt catalyst site[1]. Facile transport processes are thus essential for high-performing CLs. While the overall CL structural properties such as agglomerate size, ionomer coverage etc. is known to greatly influence various transport processes, very little is understood about the interactions in ink phase dictating the final CL structure[2,3].

In this study, we probe the impact of Pt/carbon (Pt/C) weight ratio on gas-transport resistance. We utilize a hydrogen-pump limiting-current setup to measure CL gas-transport resistance and observe that the CL resistance increases with increasing primary particle Pt loading. These results agree with previous literature data[4,5]. Additionally, the trends in CL resistance qualitatively coincide with ink zeta-potential measurements with high Pt/C weight-ratio ink samples demonstrating lower zeta-potential magnitudes. Zeta potential is a well-accepted indicator of colloidal stability, with low zeta-potential indicating lower-repulsion between ink particles and greater degree of aggregation[6]. The results hence suggest that the observed trends are due to different CL structures arising due to varying ink-interactions due to use of different primary particle loading. This hypothesis is reaffirmed from SEM imaging, where larger agglomerates and ionomer clumps are observed in CL samples fabricated using high Pt/C weight ratio primary particles.

To elucidate the observed trends further and understand ionomer-catalyst ink interactions, we study zeta potential trends with bare carbon inks while varying the ionomer content. Bare-carbon zeta potential measurements demonstrate a V-shaped curve. The behavior can be explained by segregating the plot into saturated and unsaturated regions. Ionomer is highly acidic and typically dissociates in solvent[3]. At low ionomer concentrations, the ionomer continues to adsorb on carbon surface via hydrophobic interactions resulting in increasing magnitude of zeta potential. Once the carbon surface is saturated, additional ionomer added contributes to the solvent ionic strength resulting in decreasing magnitude of zeta potential. The explanation is supported by ink pH and particle size measurements.

Acknowledgements

We would like to thank Sarah Berlinger for helpful discussions and insights. This work was funded under the Fuel Cell Performance and Durability Consortium (FC PAD) funded by the Energy Efficiency and Renewable Energy, Fuel Cell Technologies Office, of the U. S. Department of Energy under contract number DE-AC02-05CH11231.

References

1 Weber, A. Z. & Kusoglu, A. Unexplained transport resistances for low-loaded fuel-cell catalyst layers. *Journal of Materials Chemistry A* **2**, 17207-17211, doi:10.1039/c4ta02952f (2014).

2 Hatzell, K. B., Dixit, M. B., Berlinger, S. A. & Weber, A. Z. Understanding inks for porous-

electrode formation. *Journal of Materials Chemistry A* **5**, 20527-20533 (2017).

3 Berlinger, S. A., McCloskey, B. D. & Weber, A. Z. Inherent Acidity of Perfluorosulfonic Acid Ionomer Dispersions and Implications for Ink Aggregation. *The Journal of Physical Chemistry B* **122**, 7790-7796 (2018).

4 Schuler, T. *et al.* Fuel-Cell Catalyst-Layer Resistance via Hydrogen Limiting-Current Measurements. *Journal of the Electrochemical Society* **166**, F3020-F3031, doi:10.1149/2.0031907jes (2019).

5 Owejan, J. P., Owejan, J. E. & Gu, W. Impact of Platinum Loading and Catalyst Layer Structure on PEMFC Performance. *Journal of the Electrochemical Society* **160**, F824-F833, doi:10.1149/2.072308jes (2013).

6 Berlinger, S. A., McCloskey, B. D. & Weber, A. Z. Understanding Binary Interactions in Fuel-Cell Catalyst-Layer Inks. *ECS Transactions* **80**, 309 (2017).

603633 Measuring Internal Resistance of a Microbial Fuel Cell

Abstract:

We have developed a new means of quantifying the internal resistance of the microbial fuel cells. There is reason to believe that current methods for measuring the internal resistance of microbial fuel cells, such as current injection or electrochemical impedance spectroscopy (EIS), neglect a portion of the activation resistance. We propose a capacitor method as an alternative solution to this issue. Internal resistance is found by charging an external capacitor and fitting the charging curve. This method will reflect the true characteristic value of the internal resistance, because the time scale of the measurement will be consistent with the time scale of the growth of the cells. Early data analysis shows that the internal resistance obtained by the capacitor method is greater than that determined by EIS. Furthermore, for the controls performed without a biofilm present at the anode the internal resistance values found using EIS are the same as those found by the capacitor method. This suggests that the capacitor method may more accurately determine the resistance due to microbial activity.

599111 Modeling the Capacity Loss of an AGM Lead-Acid Battery

Abstract:

An absorbent glass mat (AGM) lead-acid battery is a type of valve-regulated lead-acid (VRLA) battery. An AGM lead-acid battery differs from a flooded lead-acid battery in that the sulfuric acid is absorbed in a very fine fiberglass mat, making the battery spill-proof. The leading advantages of AGM type are a higher specific power, lower internal resistance, five times faster charge speed than the flooded version, and the ability to deep cycle. AGM lead-acid batteries are found in high-end vehicles equipped with more electronic devices and making inroads into the start-stop function of cars. A validated mathematical model for the accurate prediction of the capacity loss of an AGM lead-acid battery due to the so-called aging in real-world drving conditions is necssary estimate the battery life expectancy.

In this work, a method of modeling the capacity loss of an AGM lead-acid battery under various operating conditions is presented. This work adopts a simple modeling approach by considering Ohm's law and charge conservation on the electrodes based on the simplified polarization characteristics of the electrodes. By not computing the potential distribution of electrolyte phase and the transport phenomena of ionic species, the modeling method in this work cuts down considerable computation time in comparison with rigorous electrochemical models. The model is validated by comparing the modeling results with the experimental measurements.

609065 Influence of Additives on the Performance and Failure Mechanisms of the Rechargeable Alkaline Zinc Electrode

Abstract:

Metallic zinc (Zn) is an attractive negative electrode material for rechargeable batteries because it is a non-toxic, non-flammable, earth abundant metal that is compatible with aqueous electrolytes. Zinc metal is also energy dense, exhibiting a high gravimetric discharge capacity of 820 mAh/g, a high volumetric discharge capacity of 5850 mAh/mL, and a low equilibrium potential of -1.35 V vs. mercury-mercuric oxide (Hg|HgO) in concentrated alkaline electrolyte. Rechargeable aqueous Zn alkaline batteries have thus recently been the subject of intense investigations for safe, low-cost, stationary electrochemical energy storage. Our research aims to achieve Zn depth-of-discharge (DOD) of 20% that is sustainable for 500 or more charge-discharge cycles, corresponding to an overall battery energy density of 200 Wh/L at the cell level.

However, the cycle life of the porous Zn electrode in concentrated alkaline electrolyte is observed to decrease strongly with increasing Zn DOD. For example, we recently demonstrated that metallic Zn electrodes cycled in the range of 1% to 16% Zn DOD had an exponential decrease in cycle life with Zn DOD, which was attributed mostly to the loss of active Zn material. Here, we study the effect of additives to increase the Zn DOD and cycle life of Zn electrodes. Zinc oxide (ZnO) was used as a starting material from which metallic Zn was formed in zinc-nickel (Zn-Ni) batteries. Additives were incorporated into the ZnO electrodes for the tests done on Zn-Ni batteries and these electrodes were cycled in the range of 15% to 30% Zn DOD. The potential values of the Zn electrodes vs. Hg|HgO reference electrodes were recorded to identify failure due to Zn. At 15% Zn DOD, ZnO electrodes with certain additives or combinations of additives achieved an increased cycle life compared to Zn electrodes without additives. This highlights the need to perform electrochemical and spectroscopic studies to determine the effects of the additives on the mechanisms of the Zn electrode. These additives, with further understanding, could be implemented in low-cost, energy-dense, rechargeable Zn alkaline batteries for grid-scale energy storage.

599593 Electrochemical Techniques to Detect and Quantify Li Plating after Fast Charge of Li-Ion Batteries

Abstract:

Despite electric vehicles (EVs) gaining a small foothold in the light vehicle market, long charging times and battery safety have been key limitations in EV's general acceptance. In commercial Li-ion batteries, fast charging is limited by the deposition of metallic lithium on graphite, or 'Li plating', which can cause capacity fade and catastrophic cell shorting. The ability to detect plating on-line in vehicles is critical to allow faster charging while maintaining safety standards. Analyzing the cell voltage relaxation after charge is a promising approach for in situ Li detection. Plated Li can insert into the graphite at rest, which gives a dV/dt feature unique to plating. Our work shows that voltage relaxation can be used to detect small quantities of Li plating immediately after battery charging at any charge rate, including those relevant for fast charging.

607095 Reducing Zinc Redistribution and Extending Cycle Life with Electrochemically Synthesized Zinc Oxide Anodes in Rechargeable Alkaline Batteries

Abstract:

Zinc has found its way into a host of battery systems such as zinc- air, zinc- bromine, silver-zinc, zinc- manganese dioxide and zinc-nickel batteries on account of its low toxicity, low cost, high theoretical specific capacity (820 mAh g^{-1}) and ease of handling. A preliminary economic analysis done on a large scale prismatic 1030 Wh Zn-MnO_2 battery unit has shown that if the utilization on the zinc electrode is increased (e.g., to 25% of its theoretical capacity of 820 mAh g^{-1}) then the total fabrication cost of the battery can be brought down well below $100 per KWh. In the last one year where the cost of Zn has steadily declined to about $2.5 kg^{-1}, the cost target of a under $100 per KWh battery can be achieved even with a Zn utilization of 12 - 15 %. However, due to the unpredictable behavior of metallic zinc anodes in concentrated alkaline electrolyte, its use has been greatly limited. Principal causes of cell failure have been ascribed to zinc dendrite formation leading to shorting, and to loss of capacity due to anode shape change. The problem of undesirable zinc morphology has been addressed to a great extent with the use of additives in the electrode and the electrolyte. The problem of zinc electrode shape change, however has not been fully overcome due to the movement of zincate ions in the electrolyte accompanied by electrodeposition of metallic zinc in disparate locations when cycled in concentrated alkaline electrolyte. Zinc electrode shape change is the redistribution of active material over the electrode surface as a result of cycling. This phenomenon gradually develops over several cycles resulting in reduced active electrode surface area as the zinc active material accumulates towards the bottom and other electrolyte rich regions, while severely denuding the upper regions of the electrode. To potentially mitigate this phenomenon, we investigate the cycling of an *in-situ* formed zinc-based anode comprising a highly porous electrochemically synthesized zinc oxide matrix with electrodeposited zinc particles. These anodes are formed by fully discharging 'conventional' anodes and then charging them up partly to result in anodes consisting of an electrochemically formed ZnO matrix in which the Zn particles are quite uniformly distributed and embedded. These ZnO /Zn anodes have a porosity of ~70 % which is twice that of the conventionally made metallic zinc electrode. Compared to cycling of the conventional zinc anodes at the same depth of discharge (utilization of the theoretical capacity of Zn), we find a ~70 % improvement in cycle life performance at 12 % utilization and at a rate of 10 mA cm^{-2}, without any additive. A quantitative chemical analysis of the bulk electrolyte revealed undersaturated zincate ion concentration in the electrochemically synthesized ZnO cells indicating reduced zincate movement. To gain more insight, extensive 'post mortem' measurements of Zn and ZnO masses and their distribution on both the electrochemically synthesized ZnO anodes and the conventionally made zinc anodes (the controls) were made as functions of cycle life. The experiments conducted on the electrochemically synthesized ZnO anodes showed higher retention of Zn and ZnO on the electrode surface, especially in the upper region of vertically aligned Zn anodes signifying reduced redistribution of active material compared to the controls. An Image analysis conducted on the Zn particles also revealed a narrower particle size distribution in the range of 62 - 79 μm for the electrochemically synthesized ZnO anodes compared to that of the control anodes which showed a wider particle size distribution of 51 - 96 μm. The results suggest that the electrochemically synthesized ZnO anodes provide a matrix in which the zinc particles retain their localized distribution on cycling better than the control electrodes conventionally made by binding zinc particles together, which migrate more freely on cycling.

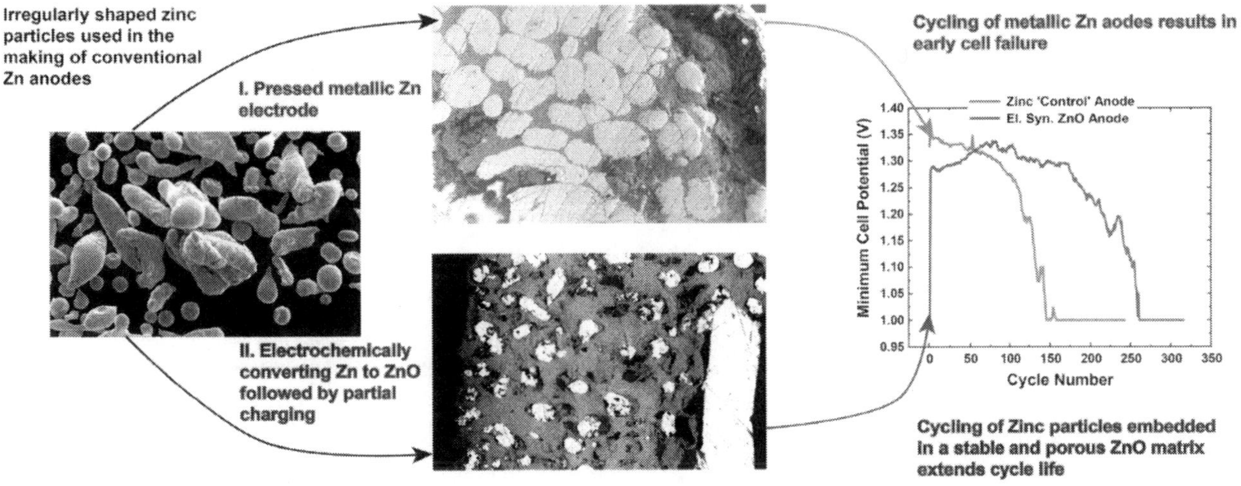
Irregularly shaped zinc particles used in the making of conventional Zn anodes
I. Pressed metallic Zn electrode
II. Electrochemically converting Zn to ZnO followed by partial charging
Cycling of metallic Zn aodes results in early cell failure
Zinc 'Control' Anode
El. Syn. ZnO Anode
Minimum Cell Potential (V)
1.40
1.35
1.30
1.25
1.20
1.15
1.10
1.05
1.00
0.95
0
50
100
150
200
250
300
350
Cycle Number
Cycling of Zinc particles embedded in a stable and porous ZnO matrix extends cycle life

609631 Persistent, Single-Polarity Energy Harvesting from Ambient Thermal Fluctuations Using a Thermal Resonance Device with Thermal Diodes

Abstract:

There is a pressing need for durable energy harvesting techniques that are not limited by intermittency, and capable of persistent and continuous operation over extended periods of time in a variety of environments. Our laboratory has identified ambient thermal fluctuations of various frequencies as potentially abundant, ubiquitous sources of such energy. We have developed a type of novel devices called thermal resonator for transient thermal energy harvesting, tested them in various environments and are working on integrating them with energy storage systems.[1, 2] In this work, we present a mathematical theory for the operation and design of a thermal resonator interfaced with optimized thermal diodes that we show is capable of much higher productivity than previous possible. This work is a significant advance over resonators that we have previously introduced using high thermal effusivity materials for a dual polarity electrical output from input temperature fluctuations. Herein, we outline the optimal design of a thermal resonance device incorporating optimal thermal diodes on its external boundaries with the environment, and we show that such a configuration is able to produce single polarity electrical generation drastically exceeding the performance of previous thermal resonators by a factor of 5. We also characterize additional thermal fluctuations in the environment, expanding the scope of resonator application. Lastly, we analyze the capability of thermal diodes available in the literature to realize such an enhancement in thermal resonance performance, and we conclude that current thermal diodes in existence could produce approximately a factor of 2 improvement in thermal resonator performance. This work points out the significant role of thermal diodes in ambient energy harvesting, and outlines important directions for next generation thermal diodes, independent of advances in thermoelectric or pyroelectric materials.

Reference:

1. L. Cottrill, A. T. Liu, Y. Kunai, V. B. Koman, A. Kaplan, S. G. Mahajan, P. W. Liu, A. R. Toland, M. S. Strano, Nat. Commun. 2018, 9, 664.
2. L. Cottrill, G. Zhang, A. T. Liu, A. Bakytbekov, K. S. Silmore, V. B. Koman, A. Shamim, M. S. Strano, Appl. Energy 2019, 235, 1514.

610622 Probing the Link between Aqueous-Phase Ce^{3+}/Ce^{4+} Structure and Redox Kinetics for Energy Storage

Abstract:

The need for energy supply to match the world's increasing population and improving quality of life has clearly been shown to be accompanied by increasing levels of CO_2 in the atmosphere. Using renewable energy sources can decrease CO_2 emissions without decreasing quality of life, but the intermittency of renewable systems such as solar and wind poses a challenge in matching energy supply to demand. The geographical limitations of pumped hydroelectric mean that for a transition to a renewable energy economy to occur, there must be new innovations in energy storage technologies. One promising energy storage technology is the redox flow battery (RFB). RFBs have the advantages of convective mass transport, long cycle lifetimes, and decoupled power and energy. However, the current densities and efficiencies of RFBs must be increased to drive down costs and make these batteries economical. Improvements to RFB performance can come from minimizing efficiency losses by fundamentally understanding the mechanism of the charge transfer in RFBs. Efficiency losses of these systems arise from overpotentials related to mass transport (also called concentration overpotential), conductivity, and kinetics. In this talk, we will focus on how understanding the structure of aqueous metal ions used in RFBs can help to understand the overpotentials for redox couples employed in aqueous RFBs.

We will briefly discuss how X-ray absorption and UV-Vis spectroscopy can be used to unravel the structure of metal ions in different acids. This redox couple is used in the well-studied vanadium RFB system. We will also show how X-ray absorption and UV-Vis can be used to study the Ce^{3+}/Ce^{4+} redox couple. Ce^{3+}/Ce^{4+} has a high positive potential[1] that makes it useful for applications for indirect oxidation, and for higher power densities for RFBs without requiring increases in current densities. We show that although Ce^{3+} is solely complexed by water, Ce^{4+} is prone to complexation with the anion, which significantly impacts the redox potential of the couple. Using our experimental methods, combined with DFT calculations, we explain the role of the electrolyte on structure and solubility. We compare the energy of Ce^{4+} complexation (determined from the change in redox potential) and the structure of the Ce ions in different electrolytes to our measured kinetic data to understand the mechanism of the reaction in aqueous phase on different electrocatalyst surfaces.

References

(1) Tucker, M. C.; Weiss, A.; Weber, A. Z. Improvement and Analysis of the Hydrogen-Cerium Redox Flow Cell. *J. Power Sources* **2016**, *327*, 591–598.

606850 Detailed Modelling and Simulation of Thermophysical Property Based Hydrogen Liquefaction Process with Exergy and Economic Analysis

Abstract:

Over the past few decades, many researchers in academia and industry have conducted to mitigate global warming. As the international convention for reducing greenhouse gas emissions was signed, the demand for hydrogen has been increasing with governments pursuing policies to solve fine dust and air quality problems [1]. In situations where large storages and transportation of hydrogen are required due to increase in demand, the development of highly efficient hydrogen liquefaction technology received much attentions [2].

In this study, a thermophyscial property based modeling and simulation approach for efficient hydrogen liquefaction with exergy and economic analysis was executed using Aspen HYSYS. For the design of the hydrogen liquefaction process, a suitable Equation of State (EOS) for accurate prediction of the hydrogen properties was investigated. Currently, NIST's REFPROP is considered as one of the most reliable EOSs [3], but seemed difficult to implement in simulation programs (e.g., Aspen HYSYS, PROII). Hence, three case studies were carried out to asssess the effect of several built-in EOSs (PR, SRK, MBWR) on the liquefaction process. The findings revealed that, MBWR was the most suitable EOS and hence was selected for this process.

In the hydrogen liquefaction process, ortho hydrogen is slowly converted to a more stable para hydrogen. To include this phenomenon in our modeling, an ortho-para conversion reactor was installed to convert the ortho hydrogen (75% of normal hydrogen) to more stable para hydrogen. Base on our modeling analsysis it was found that more than 40% of refrigerants were required when the ortho-para conversion reaction was included in the model. Furthermore, we adjusted the heat capacity parameters and equations to to take into account the conversion heat of ortho-para hydrogen so that it has the same effect as a continuous type heat exchanger that recent hydrogen liquefaction processes adopt [4]. Lastly, for the design of the heat exchanger, feasibility studies were carried out by comparing the implemented continuous type heat exchanger with the built-in batch type heat exchanger using equivalent amounts of hydrogen refrigerant in Aspen HYSYS. A minimal error of 1% was obtained confirming the possibility of using continuous heat exchangers in Aspen HYSYS with this newly developed approach. In addition, the exergy changes using different refrigerants in the pre-cooling cycle were examined and the process was validated through economic analysis.

Literature cited:

[1] K. Ohlig, L. Decker, "The Latest Development and Outlook for Hydrogen Liquefaction Technology," *Linde Kryotechnk AG.*, 2013.

[2] Alexander Alekseev, "Hydrogen Liquefaction," *Hydrogen Science and Engineering: Materials, Processes, Systems and Technology.*, Vol. 2, pp. 733–761, 2016.

[3] J.W. Leachman, R. T Jacobsen, S. G. Penoncello, and E. W. Lemmon, "Fundamental Equation of State for Parahydrogen, Normal Hydrogen, and Orthohydrogen," *J. Phys. Chem.*, Ref. Data 38, 721, 2009.

[4] Hamed Rezaie, Masoud Ziabasharhagh, Mostafa Mafi, "A review of Hydrogen Liquefaction, Current Situation and Its Future," *International Conference on Engineering&Applied Science.*, 2016

596060 Downstream Process with Countercurrent Chromatography to Recover Bioproducts from Sustainable Feedstocks

Abstract:

Separation of bioproducts produced from renewable feed streams is challenging in downstream process due to complex feed composition and low feed concentration. Current technologies such as ion exchange, adsorption, and liquid-liquid extraction are often faced with large solvent consumption, limited choice of resins, additional pretreatment and regeneration steps, and limited pH operation range. CounterCurrent Chromatography (CCC) is an emerging approach that uses at least two immiscible liquids for both mobile and stationary phase and is a potentially attractive separation method to recover bioproducts from complex process streams. Although there are multiple advantages of CCC for its liquid stationary phase in the context of natural products, there is little attention on employing CCC into downstream process for commodity bioproducts and biochemicals. This talk will discuss the feasibility of incorporating CCC into downstream process to produce high-value products from lignocellulosic feedstocks. Separation of bioproducts from lignocellulosic feedstocks such as alkaline pretreatment liquor and reductive catalytic fraction oils will be presented as demonstrations. The energy and solvent demands of the system correlated with the physicochemical properties of solvent systems, CCC operation mode, and feed composition will be discussed.

610487 Separation of Lignocellulosic Sugars from Pyrolytic Bio-Oil Using Simulated Moving Bed

Abstract:

Anhydrosugars are the major carbohydrate products of autothermal pyrolysis of lignocellulosic biomass that can be upgraded into biofuels and bio-based chemicals. Our research group has developed a fast pyrolysis process, namely autothermal pyrolysis, which can produce high yields of bio-oil by thermal depolymerization of pretreated biomass. A significant fraction of the heavy ends of bio-oil contain sugars, mainly levoglucosan, but also xylose as the major carbohydrate products. The anhydrosugars in bio-oil are acid-hydrolyzed to produce glucose, which can serve as a fermentation substrate for manufacturing common renewable fuels such as ethanol.

The bio-oil contains some phenolic monomers generated from the thermal depolymerization of lignin during the pyrolysis process. These phenolic monomers end up in the hydrolyzed, neutralized, and filtered water-soluble sugar fraction of the bio-oil due to a moderate solubility of phenol and other substituted phenolic monomers in water. The presence of phenolic species in the aqueous extraction of pyrolytic bio-oil poses significant toxicity challenges in employing fermenting microorganisms downstream and therefore must be removed from the aqueous solution of bio-oil sugars prior to fermentation. Our previous report shows effectiveness of hydrophobic polymeric resins in preferential adsorption of phenolic species compared to levoglucosan and glucose from an aqueous fraction of bio-oil.

In this work, we have developed a simulated moving bed technology (SMB) to employ continuous scale separation of sugars from phenolic species for an aqueous bio-oil fraction using the principle of adsorption chromatography. Sugars were separated from phenolics in the SMB processing a hydrolyzed, neutralized and filtered water-soluble fraction of bio-oil containing 225 g/L glucose, 10 g/L xylose and 21.1 g/L phenolics at the rate of 4 mL/min. Although it is a binary separation, the SMB is modeled to separate two classes of chemicals (sugars and phenolics) rather than two individual chemical species. The SMB uses a 3-1-1-1 configuration of six columns each having 53 cc bed volume and can operate at 0.5-8.0 bed volumes per hour (BV/h) flow rates. The columns are packed with hydrophobic resin Dowex L493 to adsorb the phenolic compounds preferentially on the resin surfaces as they are more non-polar compared to the sugars in aqueous solution. Ethanol is used to desorb the adsorbed phenolic species on the column surfaces while DI water is used to purge any adsorbed sugars on columns into a purified sugar solution. Water is also used to regenerate the columns after each cycle of feeding, desorption and purging before the next cycle of feed flow begins. The entire unit periodically switches the entry point locations of feed, purge, desorption, and regeneration steams to simulate a continuous flow of solid beds of adsorbent resin against the liquid solution of water-soluble sugars. The purified sugar product contained 154 g/L glucose and 0.8 g/L phenolics after 6 h of operation at 4 BV/h using 20 min switch interval time and 1:1 ethanol/water consumption. We have further explored the effects of switch interval time and flow rates of the streams on the separation efficiency of the SMB unit to optimize its design.

603910 Black Liquor Gasification Chemical Recovery and Combined Cycle Power at Pulp and Paper Mills Biorefinery

Abstract:

In the areas of chemicals and allied products and energy process engineering, it is of particular interest to design processes with low energy requirements (energy intensive) (preferably renewable materials, e.g., biomass) and the use of minimum raw materials or use of a byproduct as feedstock (materials intensive) and minimum number of equipment (i.e., lean manufacturing). This is the basis from which the area of process intensification (PI) arises. The main objective of PI is to propose production and manufacturing processes in plants as small as possible and with low total energy, materials, water, and other resources.

Pulp and paper are primarily made out of wood fibers originating from natural forests or pulpwood plantations. Recycled fiber and other fiber sources such as agricultural residue are also utilized and recycled fiber is becoming more commonly used in pulp and paper making. There are two significant pulping technologies available that differ greatly in terms of process, i.e., mechanical and chemical pulping. Approximately 30% of the total pulp production in European Union is from mechanical pulping while the rest is produced by means of chemical pulping. North America has major pulp and paper industry, about 21% of the total pulp produced is from mechanical pulping and rest is produced chemically.

A pulp mill that produces bleached kraft pulp generates 1.7-1.8 ton of black liquor (measured as dry content) per ton of pulp. Black liquor thus represents a potential energy source of 250-500 MW per mill. As modern kraft pulp mills have a surplus of energy, they could become key suppliers of renewable fuels in the future energy systems. Today, black liquor is the most important source of energy from biomass. It is thus of great interest to convert the primary energy in the black liquor to an energy carrier of high value. A key advantage of black liquor compared to biofuels and fossil fuels is that it is already at the mill; the handling infrastructure already exists and there are no collection and transport costs.

An extended case study is presented in which black liquor, a byproduct of wood-pulping process, is utilized to make chemicals and biofuels, in particular methanol, dimethyl ether and syngas. Different feedstock, use of byproduct and alternatives and principles of PI are commented (Das, 2020).

Reference: Das, T.K. (2020). Industrial Environmental Management: Engineering, Science and Policy, Wiley, Hoboken, NJ.

600048 From Biomass to Power Using Integrated Compact Pyrolysis, Combustor, and Stirling Engine

Abstract:

The feedstock logistic challenges, high feedstock procuring and shipping costs, high capital and operating costs, and a stringent requirement of making gasoline and diesel blends are the major barriers to commercialize biomass conversion pathways such as pyrolysis to make renewable fuels. These barriers can be overcome by developing portable technologies with a low weight, small footprint, and a high energy efficiency that can upgrade biomass to power. In this work, the viability of converting wood chips and plastic to power via integrated compact catalytic fast pyrolysis, combustor, and stirling engine technologies was determined through simulation and modeling analysis. The catalytic fast pyrolysis component consists of auger reactor followed by a catalytic upgrading reactor and the air-cooled condenser that upgrades wood chips and plastic to a high energy dense fuel. The ASPEN simulations followed by sizing analysis were performed to determine the size, weight, and net energy efficiency of producing 42.5-Watt equivalent of power generation from wood chips and plastic using the integrated portable technology. The impact of catalytic fast pyrolysis oil yields, oil quality, condenser type and its efficiency, type of fuel combustor, and combustor efficiency on the size, weight, and net energy efficiency of upgrading biomass to power was determined.

602795 Techno-Economic Analysis and Life-Cycle Assessment of Emerging Technologies for Bioprocessing Separations

Abstract:

Limited availability, rising costs, and environmental concerns about fossil fuels have generated considerable interest in finding alternative, renewable sources including biomass, which can be converted into a number of biofuels and bioproducts. In comparison to petroleum-based products, high processing costs, mainly associated with bioprocessing separations, limit widespread implementation of biofuels and bioproducts. Bioprocess-related separations are also complicated, regardless of the conversion pathway, due to the dilute nature of products and the chemically complex mixtures that result from biomass deconstruction. In many cases, bioprocessing separation approaches lack a technology baseline, or definition of the state of technology (SOT). This work focuses on evaluating the technology readiness of novel separations technologies for the conversion of biomass into biofuels and bioproducts and will address three key topics: 1) SOT description, 2) economic viability analysis of both SOT and innovative separations, and 3) environmental impact assessment of both SOT and novel processes. The current SOT for lignin valorization, dilute carbon recovery, and impurity removal is used to identify potential opportunities for improvement and provide a baseline for comparison with the emerging technologies. Detailed techno-economic analysis (TEA) and life cycle assessment (LCA) are applied to understand the key drivers and challenges related to the economic feasibility and environmental impacts, respectively.

607169 Promoted Effect of Maleic Acid As an Additive on Thermal Dehydrogenation of Ammonia Borane

Abstract:

Renewable energy has been attracting keen interest with growing concerns over depletion of fossil fuel and climate change. Hydrogen is considered as an alternative energy carrier due to its high calorific value and no greenhouse gas emission. Among the storage technology of hydrogen, ammonia borane (AB, NH_3BH_3) is an attractive candidate owing to its high hydrogen content (19.6%), decomposition under the moderate condition, and stability at room temperature. However, relatively high operating temperature, long-induction period, low H_2 yield, and slow kinetics remain challenges for thermal dehydrogenation of AB. In the present work, the effect of maleic acid as an additive to promote dehydrogenation of AB was investigated for various operating parameters such as AB-maleic acid ratio, temperature, and pressure. Maleic acid has been shown to plays a role in lowering the onset temperature (65 ^{o}C), while increasing the hydrogen release kinetics of AB. In addition, TPD/MS and FR-IR analysis were performed to understand the reaction mechanisms.

610252 The Pre-Acidification Gas Impact on Upgrading the Biogas Produced in Expanded Granular Sludge Bed Reactor

Abstract:

Two-stage anaerobic reactors are being widely used in the organic waste management industry. In these reactors, up to one-third of the chemical oxygen demand (COD) content is naturally preacidified in a first stage pre-acidification (PA) and then fed to a second stage digester for conversion to methane. Traditionally, all the generated gases from the PA tank will be vented to the atmosphere. Hydrogen and carbon dioxide are the main gases generated in the PA tank. A pilot scale of two-stage anaerobic expanded granular sludge bed reactor was fabricated and used to investigate the impact of the PA gas injection into the second stage. The gas from the PA reactor was captured and stored in storage tank. The tests were run under two temperature ranges and five organic loading rates (~2, 3, 4, 5, and 6 g COD/L.day). For mesophilic range, the biogas production and energy yield increased by 10-90% and 40-130%, respectively, from without PA gas injection case compared to with PA injection case. For thermophilic range, the biogas production and energy yield increased by 12-40% and 90-140%, respectively, from without PA gas injection compared

to with PA injection case. For each OLR, the gas production and energy yield were 90 to 160% more in thermophilic range than the mesophilic range for the cases with and without the PA gas injection. This implies that higher temperature range has a significant impact on energy yield in a digester. One of the important finding was the amount of the PA gas injected into the EGSB reactor should be less than 50% of the theoretical calculated hydrogen gas based on ethanol substrate assumption.

611806 On Process Intensification through Storage Reactors:a Case Study on Methane Steam Reforming

Abstract:

In this work, the novel storage reactor (SR) process is introduced. A SR consists of two physically distinct domains, designated as the reactor domain and the storage domain, which are allowed to communicate with each other through a semipermeable boundary. It is envisioned that the SR process is dynamic (periodic), enabling the loading and unloading of the storage domain through Partial Pressure Swing Operation (PPSO). In this work, a first principle SR model is developed that quantifies SR dynamic behavior and a illustrative case study on Hydrogen production through Steam Methane Reforming (SMR) is carried out. Through parametric studies on reactor and catalyst structural characteristics parameters it is shown that the SR outperforms an equivalent steady-state reactor.

595543 Co-Production of Hydrogen and Butanol By Fermentation with the Microorganism Cells Immobilized on Bagasse

Abstract:

The production of biofuels by fermentation is widely concerned owing to its mild operation conditions and broad substrate. As a kind of biofuel, bioethanol can be added into the gasoline and the mixture has been used in industry. Apart from bioethanol, biobutanol is also a good addition owing to its higher calorific value, lower vapor pressure, less corrosive to equipment and better applicability in the existing car engine. Biohydrogen is always regarded as one promising fuel for its high calorific value and cleaning burning waste (only water). Reports showed that the bioproduction of hydrogen and butanol can be achieved together by fermentation. In this study, the *Clostridium acetobutylicum* was used to generate hydrogen and butanol. At the very beginning of the fermentation, the cells were immobilized on carriers by surface adsorption and then the carriers with cells were used in different mode of fermentation such as flask batch fermentation, flask repeated cycle fermentation, repeated cycle fermentation in the fixed bed, coupled system of continuous fermentation and in situ membrane separation module and coupled system of repeated cycle continuous fermentation and in situ membrane separation module to explore the corresponding fermentation performance.

Firstly, three kinds of materials including activated carbon, bagasse and brick were chosen to be used as carriers to attach cells. The optimal carrier should be determined according to the fermentation performance and attachment capacity. SEM pictures showed that cells were apparently attached on all of the three kinds of materials' surface. Biofilms composed of cells were also formed on several part of the carriers' surface (cell layer was 3 micrometer thick on the activated carbon surface) and the cells on the carrier surface were orientated analogously as shown in Fig. a and b. Experiment showed that the butanol yield were with the highest value 0.22 g/g when the bagasse was used for attaching. The highest hydrogen yield 1.81 mol/mol was achieved when the brick was used as the carrier. It seems that different kinds of carrier has different effect on the fermentation performance. Then, to explore the reuse feasibility of the immobilized cells, repeated cycle fermentation in flasks was conducted. There were total 7 cycles almost 312 h in this experiment. Results showed that the fixed cells can indeed be reused for several times with a relatively high cell concentration (the highest ratio of cell dry weight and bagasse was 0.25). But with the fermentation proceeding, the cell morphology and fermentation performance were gradually changed. Cell morphology changed from the short-rod existed in the 1st cycle to the long-chain (cells united one by one) appeared in the 7th cycle. Fermentation performances got worsen step by step. While, in the first 4 cycles, about 211 h fermentation, the hydrogen production and yield were hardly decreased and they were maintained at 5.8-6 L/L and 1.5 mol/mol as it was shown in Fig. c. But the butanol concentration and yield were decreased gradually from 6.7 g/Land 0.22 g/g to 2.5 g/L and 0.12 g/g, respectively shown in Fig. d. The main solvents metabolites produced by the cells were acetone, butanol and ethanol. With the fermentation proceeding, the proportion of butanol in the total solvent-ABE was changed and the mass ratio of A:B:E was changed from 3:6:1 to 1:3:9 showing that the distribution of carbon in the metabolites was changed during long time fermentation.

Experiment in the 5 L fermenter was conducted to see if the fermentation with immobilized cells would work well after scaling up. After 4 cycles, 228 h fermentation, results shown in Fig. e and f illustrated that the maximum production and yield of hydrogen and butanol were 11.7 L/L and 1.81

mol/mol and 10.4 g/L and 0.18 g/g, respectively. Fermentation performances in the scaled up reactor were similar with that of the fermentation in flasks identifying the feasibility of scaling up. In the conventional fermentation, the butanol concentration is hard to exceed 20 g/L and generally stays at 12 g/L owing to the inhibition of high-concentration butanol (usually 13-14 g/L) to cells. To remove the inhibition and enhance the final butanol production, *in situ* membrane separation was introduced in this experiment. The membrane used in this process was made in the laboratory and it's a kind of composite membrane with the PA used as the substrate and PDMS used as the separation film. Experiment showed that the highest accumulated butanol concentration was 34 g/L which was three times as much as that of the traditional batch fermentation. The accumulated hydrogen production was 33 L/L, which was also three times as much as that of the batch fermentation 12 L/L. These information indicated that the inhibition was efficiently alleviated by separating the butanol from broth. During the whole separation process, there was no obvious membrane fouling appearing and the separation performance of the composite membrane was relatively stable as they were shown in Fig. 4. The average total flux and butanol separation factor were around 500 $g/m^2/h$ and 12, respectively. High butanol concentration (116 g/L) with stratification was obtained on the membrane downstream during the fermentation process. To study the feasibility of the cell reuse in this coupled system, repeated cycle fermentation was also conducted. There were two cycles fermentation, about 420 h. The accumulated butanol concentration was 27 g/L in the 2nd cycle which was slightly lower than 34 g/L but still higher than that of batch fermentation. But the hydrogen production was just 19 L/L which was much lower than 33 L/L while it was also higher than that of batch fermentation. Although the overall fermentation performance was not ideal, the repeated cycle fermentation mode saved operating time because there was no need to disassembly and wash fermentative equipment between two cycles fermentation but it is a must in the conventional batch fermentation process.

Different fermentation performance was obtained when the fermentation process was different. But, overall, the yield of hydrogen and butanol was similar indicating the metabolic pathway of the cells stayed stable. The method of cell immobilization used in this study achieved the reuse of high-concentration cells and it saved the equipment-washing time and the cell lag phase to some degree. *In situ* membrane separation was proved to be a very promising method to enhance the butanol production by alleviating the inhibition from butanol to cells.

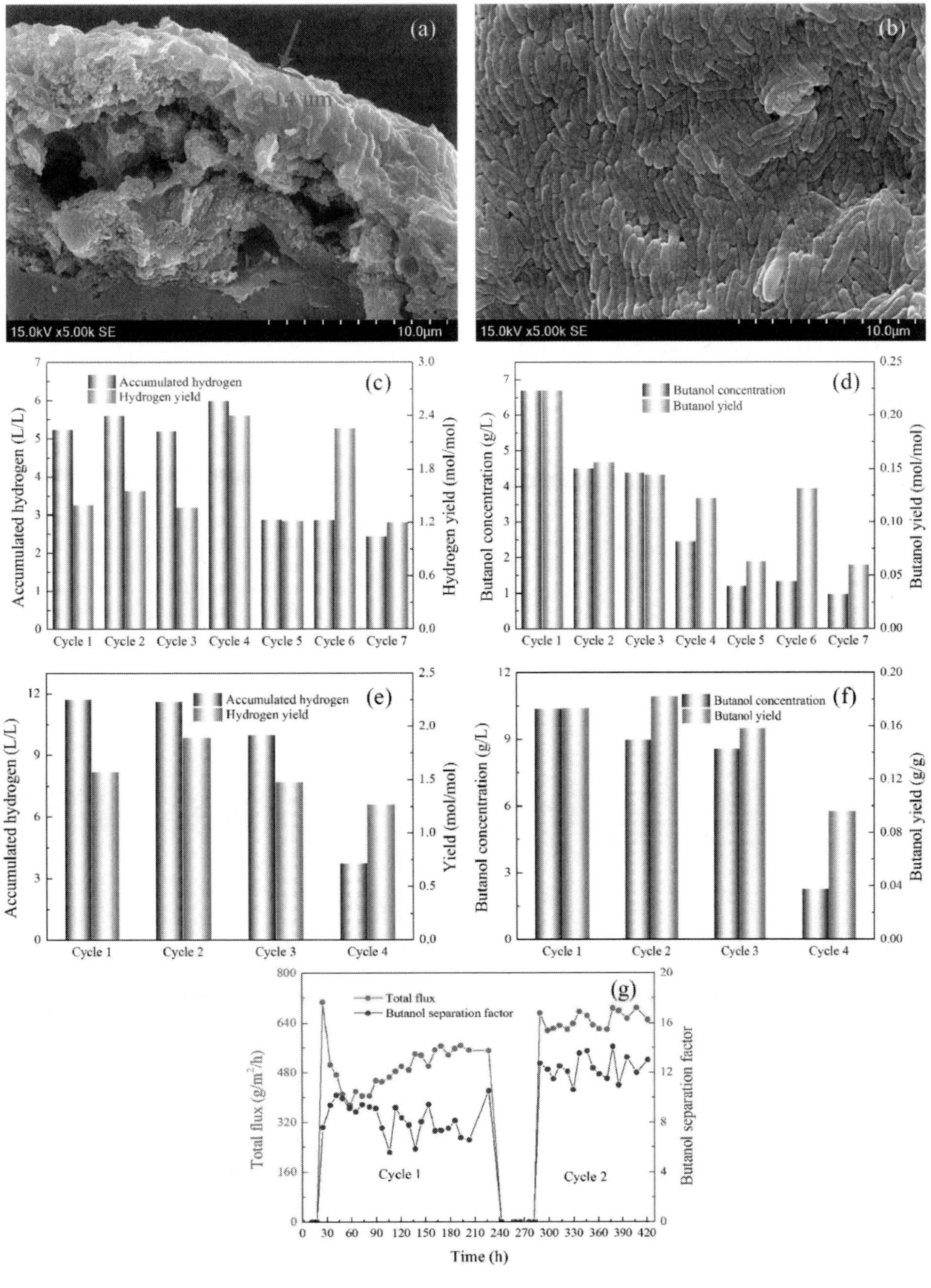
(a)
15.0kV x5.00k SE
10.0μm
(b)
15.0kV x5.00k SE
10.0μm
(c)
Accumulated hydrogen
Hydrogen yield
Accumulated hydrogen (L/L)
Hydrogen yield (mol/mol)
Cycle 1
Cycle 2
Cycle 3
Cycle 4
Cycle 5
Cycle 6
Cycle 7
(d)
Butanol concentration
Butanol yield
Butanol concentration (g/L)
Butanol yield (mol/mol)
(e)
Accumulated hydrogen
Hydrogen yield
Accumulated hydrogen (L/L)
Yield (mol/mol)
(f)
Butanol concentration
Butanol yield
Butanol concentration (g/L)
Butanol yield (g/g)
(g)
Total flux
Butanol separation factor
Total flux (g/m²/h)
Butanol separation factor
Cycle 1
Cycle 2
Time (h)

609061 Solar Thermochemical H2O/CO2 Splitting Cycle Using Ferrites

Abstract:

Solar H_2/CO generation via thermochemical water splitting (WS)/ CO2 splitting (CS) cycle was scrutinized by several research teams by employing zinc oxide, tin oxide, iron oxide, ceria and doped ceria, perovskites, and ferrites as the redox active materials. In case of ferrites, Ni-based, Zn-based, and Co-based ferrites, Cu-based ferrites have also shown promising results. This study deals with the estimation of the solar-to-fuel energy conversion efficiency of the Cu-ferrite based WS (CFWS) cycle by performing a detailed thermodynamic equilibrium and efficiency analysis. This analysis was conducted by utilizing the thermodynamic properties obtained from HSC Chemistry software. Various process parameters such as solar energy required, energy released during the H2O/CO2 splitting cycle, amount of energy that can be recuperated and others were determined by drifting the partial pressure of O2, TR temperature and WS temperature. The obtained results were compared with the previously, thermodynamically, investigated Ni-ferrite based WS (NFWS) cycle. As per the literature review, the thermodynamic evaluation of the CFWS cycle was not yet conducted.

612524 Advanced Carbon Capture System for Power Plants

Abstract:

TDA Research, Inc (TDA) is developing a low cost, high capacity CO_2 adsorbent and demonstrating its technical and economic viability for CO_2 capture from coal-fired power plants. TDA uses an advanced physical adsorbent to selectively remove CO_2 from the flue gas. The sorbent exhibits a much higher affinity for CO_2 than N_2, H_2O or O_2. The sorbent binds CO_2 more strongly than common adsorbents, providing the chemical potential needed to remove the CO_2. However, because CO_2 does not form a true covalent bond with the surface sites, the regeneration can be carried out with only a very small energy input (4.9 kcal per mol of CO_2) compared to amines (14.2 kcal/mol CO_2). We demonstrated the sorbent's efficacy through multiple cycles for removing CO_2 from coal derived flue gas and the sorbent achieved a stable capacity. The presence of acid gases such as SO_2 and NOx (80 and 200 ppmv, respectively) and water vapor up to 15% vol. caused no adverse effect on the CO_2 capacity. The regeneration of the sorbent and the recovery of CO_2 and its pressurization can be achieved by several approaches, including temperature and vacuum swing. A detailed techno-economic analysis using Aspen Plus™ software is conducted in collaboration with University of California Irvine to estimate the impact of the TDA's CO_2 capture system on plant efficiency and cost of electricity. Our system analysis showed that TDA's CO_2 capture process will reduce plant efficiency only by 19.4% (compared to 31.2% reduction for amine scrubbers) while providing 90% capture and the increase in 1st year levelized cost of electricity (COE) is 18.4% point lower than that of amines. We will present the latest results from the DOE sponsored work at the AIChE meeting.

THE USE OF DIGITAL TOOLS IN THE SEARCH FOR NEW SOLVENTS TO CAPTURE ACID GASES LIKE CO_2

Frédérick Jean-Marie de Meyer, TOTAL S.A., Paris, France

Abstract

Molecular simulations (Molecular Dynamics, Monte Carlo) and chemoinformatics are used quite extensively to identify new adsorbents for acid gases and particularly for CO_2. The use of those digital tools is far less common (MD, MC) or almost non-existent (chemoinformatics) in the search for new solvents (absorption) to capture acid gases. This is mainly due to the complexity of the absorption process. We have applied quantum chemistry, molecular simulations (MD, MC), and chemoinformatics to predict the absorption capacity of solvents and to identify solvents with a superior performance (solvent screening). The results were compared with experiments. This allows us to evaluate the different approaches. We present the strengths and weaknesses of each digital tool.

Article

Introduction

The removal of acid gases from both natural gas and flue gases is a key chemical process which is expected to become even more important in the future. It seems almost impossible to meet the climate goal without the capture and storage of CO_2 (CCS), mainly from flue gases (anthropogenic CO_2), but to a lower extent also from natural gas (native CO_2). The most common and mature industrial method to remove acid gases from both the low-pressure flue gases and the high-pressure natural gas is absorption into a liquid solvent, particularly aqueous alkanolamine solutions. The captured CO_2 can be stored (e.g., Enhanced Oil Recovery), valorized, or is emitted to the atmosphere. If H_2S is present in the natural gas, it is also removed with the same Acid Gas Removal Unit (AGRU) and typically further converted into elemental sulfur via the Clauss process[1].

A typical AGRU consists of an absorption column, tray, or packed, where the acid gas is in countercurrent contact with the aqueous amine solvent. The acid gas molecules are first physically absorbed in the solvent. Subsequently, the acid gases in solution react with the aqueous amine via acid-base reactions and are converted into non-volatile ions. The solvent loaded with acid gas is sent to a regenerator where steam is used to strip the solvent from its acid gas. The lean solvent is recycled to the absorption column, while the acid gas is separated from the water. An example of an AGRU with the typical operating conditions is given in Figure 1 for natural gas treatment.

New solvents are currently also being developed to reduce the loss (e.g. emissions) of potentially harmful solvent degradation products, to increase the absorption capacity and absorption rate the of acid gases and to increase the selectivity of the absorption. But, the main drawback of CO_2 capture with an aqueous alkanolamine solution is the energy consumption of the regeneration due to the use of steam. A significant amount of research effort has been put into the development of less energy-intense CO_2 capture technologies.

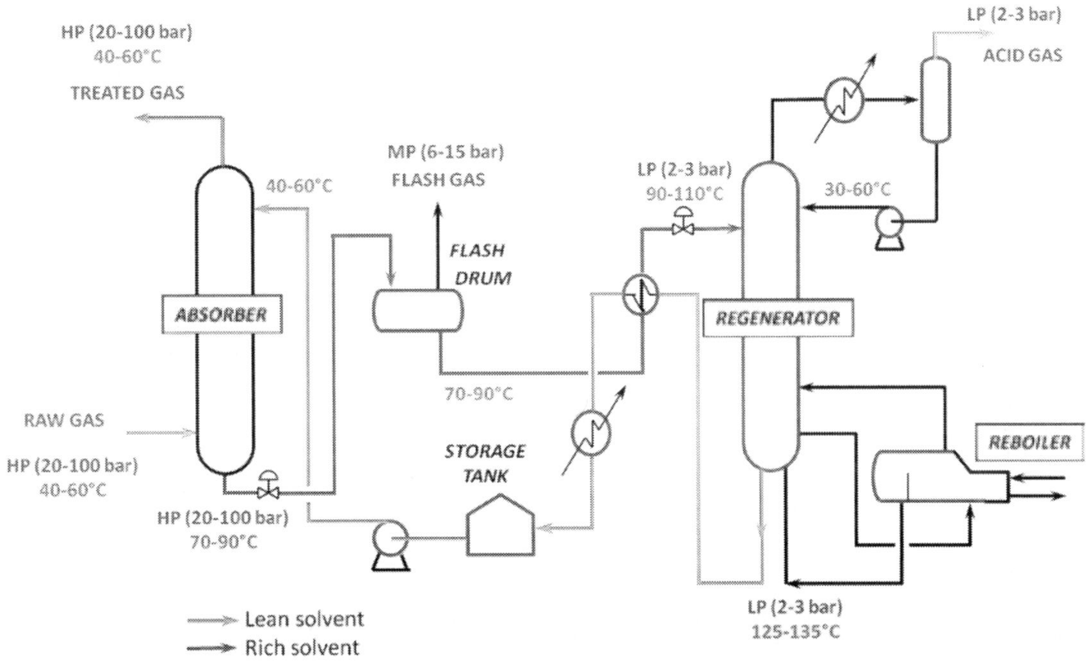

Figure 1: Process flow diagram of an AGRU with operating conditions typical for natural gas.

Part of these technologies is not based on solvents, like adsorption-based technology, membranes, and cryogenic distillation, and part is based on solvents like advanced strippers, water-lean solvents, etc. As none of the non-solvent based technologies seem to be technically or economically ready for large scale industrial deployment, the solvent-based technology will likely continue to dominate the market in a near future.

Figure 2 illustrates the physics and chemistry involved in the absorption of H_2S and CO_2 by an aqueous tertiary amine. The acid gases are first physically absorbed in the amine base solvent. Subsequently, they are involved in acid-base reactions and transformed into non-volatile ions. Changing the solvent thus changes the physical absorption, the chemical reactions, the ion solvation effects, mass transfer (diffusion, viscosity, etc.) the pH, desorption properties, etc. The absorption process is thus quite complicated. Moreover, despite the considerable past efforts to further improve and optimize acid gas capture with solvents, several fundamental physical and chemical aspects of the process remain not clear, and this strongly inhibits further progress.

In the search for new adsorbents, digital tools like quantum chemistry, molecular simulations and chemoinformatics are quite extensively used[2]. The use of those digital tools is far less common (molecular simulations) or almost non-existent (chemoinformatics) in the search for new solvents to capture acid gases. This is mainly due to the higher degree of complexity of the absorption process.

We have applied quantum chemistry, molecular simulations (MD, MC), and chemoinformatics to predict the absorption capacity of solvents and to identify solvents with a superior performance (solvent screening). This allows us to evaluate the different approaches. The aim of this paper, which mainly addresses engineers and scientists who are not familiar with using these digital tools, is to present the strengths and weaknesses of each digital tool with simple examples. This is done by comparing the efforts and accuracies of the different approaches to model and predict the physical CO_2 solubility.

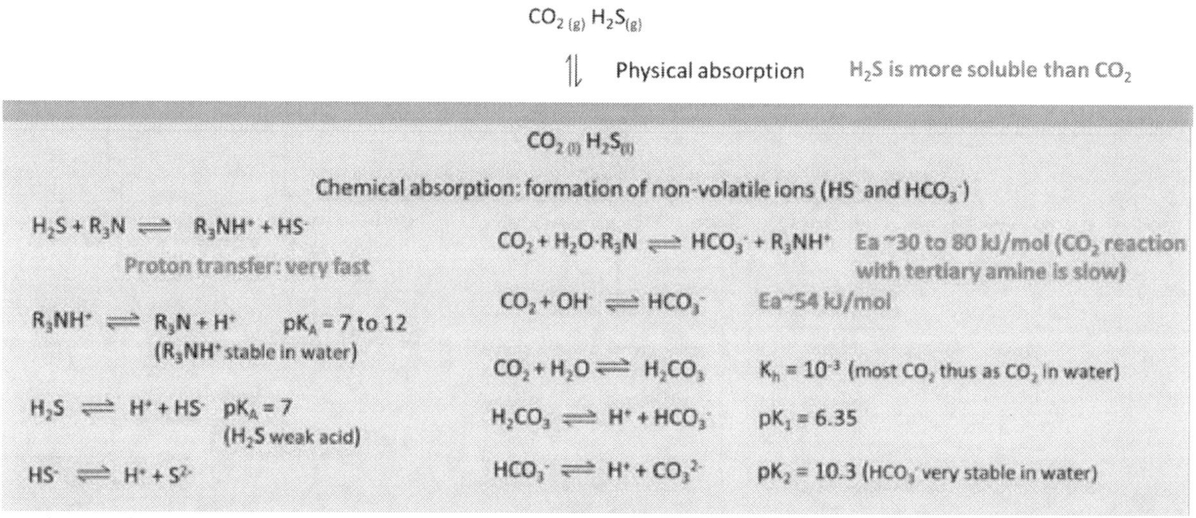

Figure 2: Example of the physics and chemistry involved in the absorption of acid gases by aqueous tertiary amines.

In-house group contribution methods

As a first approach, to gain some fundamental insight and to evaluate the more advanced digital tools, simple group contribution methods to predict absorption-related properties have been developed in-house.

There exist many ways to thermodynamically describe the physical solubility of a gas in a solvent. One is to use the mole fraction x_i of solute i in a solvent when the gas and liquid phase are in thermodynamic equilibrium. The advantage of using the mole fraction is that it is directly available from experiments. In the limit of infinite dilution it is experimentally observed that x_i is proportional to the partial pressure of component i in the gas phase. The proportionality constant is called the Henri constant:

$$P_i = H_i x_i$$

Starting from a VLE expression and assuming the gas phase to be ideal and the Poynting factor to be negligible one can show that for a solute i, at a given temperature T, the Henri constant relates to the saturation pressure:

$$H_i = P_i^{sat} \gamma_i$$

with

- P_i^{sat} the saturation pressure of component i at the working temperature T
- γ_i the activity coefficient of solute i in the solvent.

The activity coefficient is related to the excess partial molar Gibbs free energy:

$$ln\gamma_i = \frac{\overline{G_i^E}}{RT}$$

The excess (with respect to ideal solution) partial molar Gibbs free energy of the solute i in the solution is a measure for the change in excess Gibbs free energy due to the addition of a solute to the solvent.

For an ideal solution (equal strength of self- and cross interactions between components) $\overline{G_i^E}$ is by definition zero. The activity coefficient is thus one, and $H_i = P_i^{sat}$ and, $P_i = P_i^{sat} x_i$ (Raoult's law). In other words, the solubility of i depends directly on its saturation pressure at the working temperature. The higher the saturation pressure of the solute, the lower its solubility.

For a non-ideal solution with an attractive interaction between solutes and solvent the $\overline{G_i^E}$ is negative and the activity coefficient $0 < \gamma_i < 1$, resulting in higher solubilities.

For a non-ideal solution with an unfavorable interaction between solutes and solvent the $\overline{G_i^E}$ is positive and the activity coefficient $\gamma_i > 1$, resulting in lower solubilities.

Thus, the solubility is not only determined by the solvent-solvent and solvent-solute interactions (the partial molar excess Gibbs free energy or the activity coefficient), which are relatively difficult to describe and understand, but also by the saturation pressure of the solute.

The saturation pressure is even the most important effect and it explains the main observations. As an example, for CO_2, H_2S or CH_4, the saturation pressure increases significantly with temperature and, as a result, this will negatively impact the solubility. Also, the ranking in saturation pressure of the solutes is directly related to the ranking of solubility of the solutes in any given solvent. For example: Psat(CH_4)>Psat(CO_2)> Psat(H_2S). This will push all solvents towards a higher solubility of H_2S than CO_2 and CH_4 which is indeed experimentally observed, indicating that intermolecular interactions are of secondary importance.

There exists much experimental data on the solubility of CO_2 in physical solvents at different pressures and temperatures. Part of the data have been collected and subjected to critical analysis by IUPAC. From various papers we have collected/calculated the Henry coefficient (in MPa), at 20 and at 50°C. We have developed a very simple group contribution method to predict the solubility of CO_2 at 20°C and at 50°C. The Henry coefficient of CO_2 in a solvent can be estimated with the formula:

Henry CO_2 = X0 + X1*number of primary or secondary amines + X2*number of tertiary amines + X3*number of amides + X4*number of OHs + X5*number of ethers (-O-) + X6*number of C + X7*number of C=O + X8*number of S=O + X9*number of rings + X10*number of Phosphates + X11* number of Cl + X12*number of Br, I. The values of Xi are obtained by fitting against the experimental data and are given in Table 1.

Acetates, esters, aldehydes, etc. are considered a combination of –O-, -OH, C=O, etc. No distinction is made between aromatic and non-aromatic rings or between alkanes and alkenes.

		20°C	50°C
	Variable X0	6.254	11.256
Prim/Sec N	Variable X 1	9.470	N.A.
Ter N	Variable X 2	0.736	0.189
Amide	Variable X 3	-1.824	-1.191
OH	Variable X 4	10.182	13.032
O	Variable X 5	-1.663	-2.172
C	Variable X 6	-0.049	-0.215
C=O	Variable X 7	-2.198	-0.658
S=O	Variable X 8	4.117	-0.556
ring	Variable X 9	3.390	3.462
Ph	Variable X 10	-2.978	-4.347
Cl	Variable X 11	0.785	1.380
I/Br	Variable X 12	2.950	4.620

Table 1: values Xi to be used for GC method to predict CO_2 solubility.

Despite its simplicity, the variables X0 obtained are very close to the experimental saturation pressures of CO_2 at 20°C and at 50°C, which are 5.72 Mpa and 11 MPa, respectively. In an ideal solution those saturation pressures should be the Henry constants. Thus, the group contribution method starts with assuming an ideal solution and adds the non-idealities of the different groups. The variables X1/X0 to X12/X0 are basically the contributions of every structural group to the overall activity coefficient. One can immediately see that alcohol groups significantly decrease the CO_2 solubility, while for example carbons, ethers, amides and phosphor groups increase the CO_2 solubility. Again, the method is extremely simple and fast, but allows to get some first basic important insights and is sufficiently accurate for screening of a large list of solvents if one is only interested in identifying the extremes (e.g. highest CO_2 solubility). The relative error is estimated at 39% (20°C) and 34% (50°C). A similar method has been developed for H_2S, CH_4, etc.

Chemoinformatics

Chemoinformatics[3] are basically very similar to the group contribution approach. The different steps are explained in the Figure 3 below. As an illustration, one can apply the chemoinformatics approach for the physical solubility of CO_2 in solvents. In a first stage all the data on CO_2 solubility in physical solvents is collected. Subsequently descriptors (e.g. different functional groups or quantum chemical properties, etc.) are generated to describe the solvents. During the so-called "machine learning" step a quantitative relationship between the solvent structure (the descriptors) and the property (the solubility of CO_2) is obtained. The resulting model is the QSAR: Quantitative Structure Activity Relationship. In the next steps one can use this predictive model to screen a list of solvents for their capacity to dissolve CO_2.

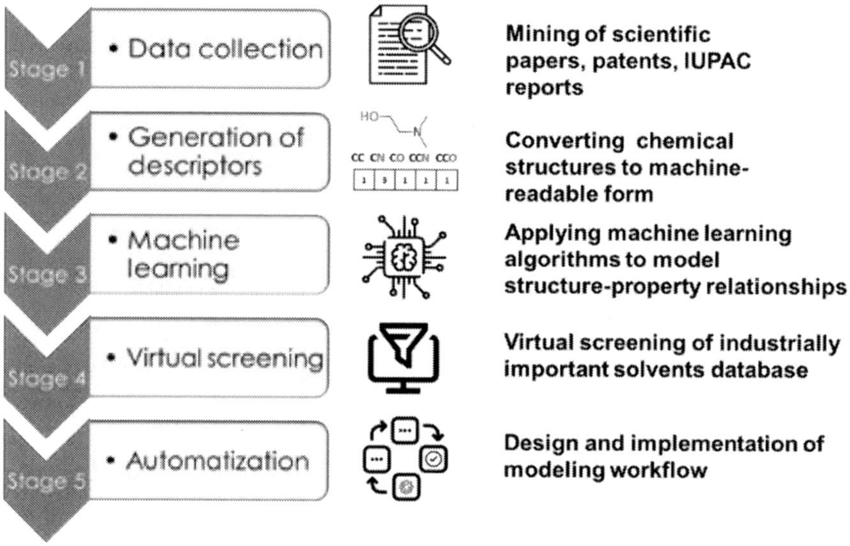

Figure 3: The 5 stages of chemoinformatics applied to solvent screening.
(courtesy of Dr. Alexey Orlov, University of Strasbourg)

Applying the first 4 stages to the solubility of CO_2 and CH_4 in various solvents typically results in the graph as shown in Figure 4. The Figure 4 shows the CO_2/CH_4 selectivity versus the solubility of CO_2. The solubility of CO2 is expressed as the Kuenen coefficient, which is directly related to mole fraction solubilities. The Kuenen coefficient is the volume of gas dissolved at normal conditions per unit of solvent mass. It is often used in the industry because it directly relates absorption capacity to mass of solvent (and thus the cost). Moreover, the mole fraction solubility of a solvent might be high simply because it is a large but heavy molecule (see group contribution methods). Using the Kuenen solubility avoids this issue.

In Figure 4 the blue dots represent the experimental data which have been used to build and validate the chemoinformatics model. The orange dots are the results of the screening of a large data base containing several thousand industrial solvents. The relative errors of these models is around 25%, which is, as expected, lower than the simple group contribution methods.

However, Figure 4 also nicely illustrates some drawbacks of chemoinformatics. The predicted orange dots all fall in between the experimental blue dots, indicating that chemoinformatics models tend to interpolate. Moreover, when they extrapolate the errors are larger. Another disadvantage of chemoinformatics is that the models tend to underperform for rare chemical structures for which there is a lack of data. Those structures are often the most interesting ones, e.g. those with the best solubilities. Consequently, with chemoinformatics it is not obvious to identify molecules performing better than the best experimental solvents. To do so one needs to first collect experimental data for structures similar to the best performing experimental molecules and then, working step by step, generate and screen their structural analogues, followed by another experimental campaign around the best performing ones and so on.

Thus, although chemoinformatics is fast and can be relatively accurate, it mainly screens within a well-known chemical space defined by the database. To identify much better performing molecules one needs to combine chemoinformatics with experimental campaigns.

Obviously, machine learning can only work if there is a clear relationship between structure and property. If very subtle changes in structure result in significantly different properties it will be very challenging for chemoinformatics to capture this.

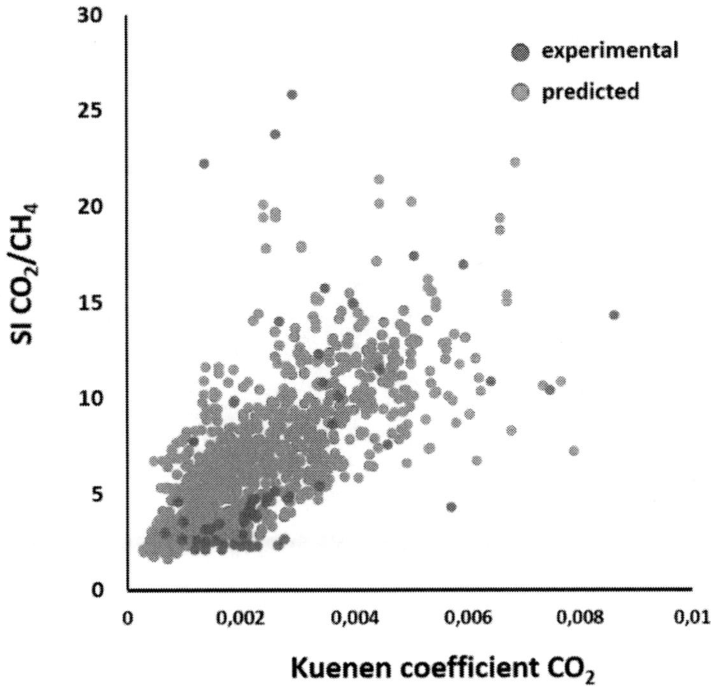

Figure 4: Solvent selectivity of CO_2 versus CH_4 as a function of CO_2 solubility expressed as the Kuenen coefficient. Blue dots are experimental datapoints, while orange dots are the results of a screening of an industrial solvent database (courtesy of Dr. Alexey Orlov, University of Strasbourg)

Molecular Simulations

To develop a group contribution method or a decent chemoinformatics model, a structure-property relationship should exist and one typically needs a large database containing at least around 100 datapoints. This is not always available. As an example, to predict the CO_2 absorption properties of chemical solvents like aqueous amines, there is simply not enough experimental data to use those methods. Despite this, different research groups have tried, with negative results (inaccurate models). At this point molecular simulations (quantum chemistry, Monte Carlo Simulations, Molecular Dynamics) might be useful[4].

This is a very broad area of research and for the sake of simplicity and coherence the discussion is limited to the solubility of CO_2 in physical solvents. Figure 5 compares the experimental temperature dependence of the Henry coefficients of CO_2 in various physical solvents with those obtained using a Monte Carlo simulation method. It should be noted that Monte Carlo simulations uses force fields which have been parametrized using experimental solvent properties (e.g. the density). Although the method qualitatively correctly predicts the solubility of CO_2, the relative error can sometimes be large, specifically for solvents which form Hydrogen bonds, like water. It is well known that this effect is challenging to capture with molecular simulations.

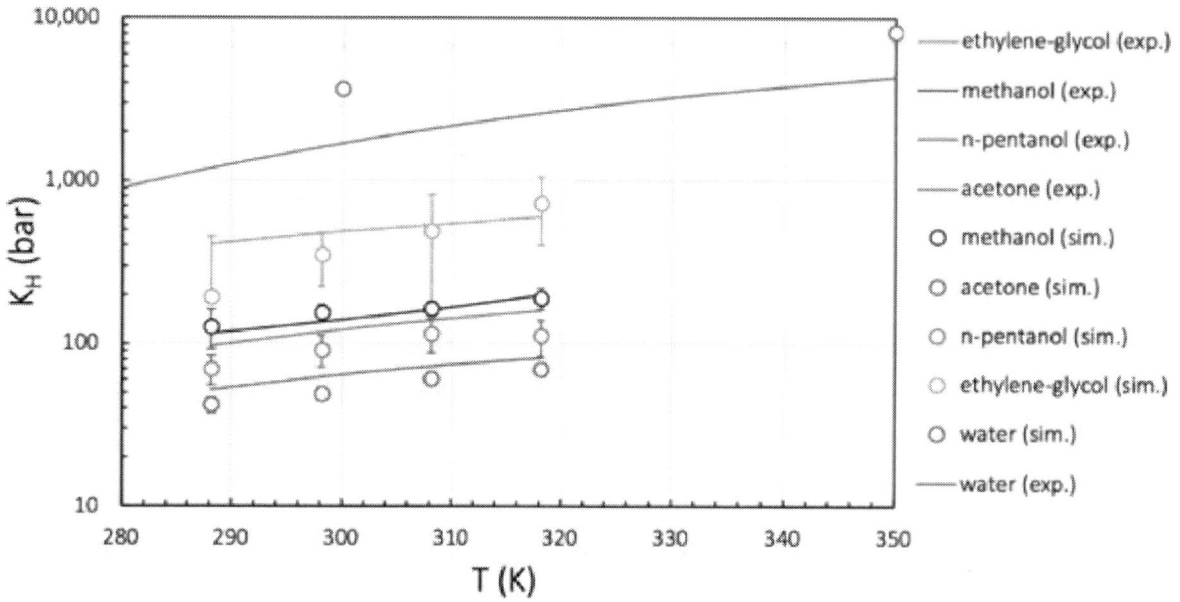

Figure 5: Henry coefficients of CO_2 in various solvents and as function of temperature. Comparison of experimental values and Monte Carlo simulation results. Courtesy of Dr. Marianna Yiannourakou from Materials Design.

Alternatively, one could, in theory, calculate the Henry coefficient from the Gibbs solvation free energy obtained by Density Functional Theory. The results are shown in Figure 6. The Gibbs solvation free energy has been obtained with the SMD implicit solvation method[5]. The accuracy is very poor, even after a systematic correction of the data, the relative error is more than 100%, meaning the results are basically random. It is thus not recommended to use this type of methods.

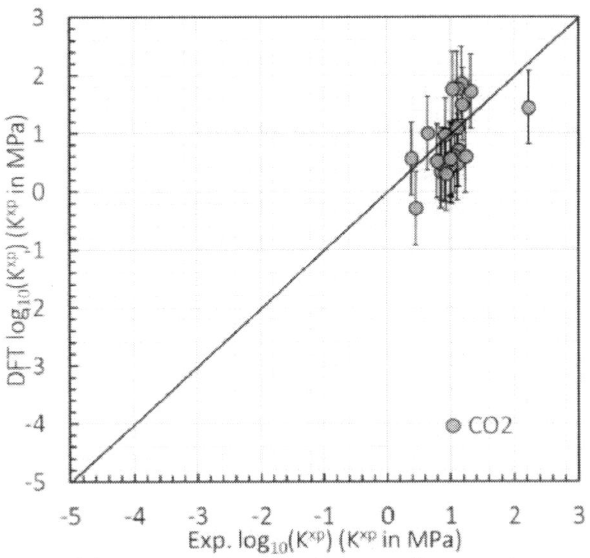

Figure 6: Henry coefficients of CO_2 in various solvents. Comparison of experimental values and Density Functional Theory-based results. Courtesy of Dr. Xavier Rozanska from Materials Design.

Conclusions & future perspectives

This paper briefly presents the main digital tools that can be used in the search for new solvents. Each tool is applied to obtain the physical solubility of CO_2 in various solvents. From this simple example general conclusions can be drawn.

If sufficient reliable experimental data of a given property are available it is worth trying to build a group contribution method and to use it for a first screening. This is fast (several days/weeks), not very complicated, and results in useful insights. If one needs a more accurate model, a chemoinformatics tool can be developed. However, this takes significantly more time (several months to a year) and specific knowledge (many tools are freely available on the internet) and it is a real challenge to use such models to identify much better performing solvents without extra experimental work. Obviously, group contribution methods and machine learning only work if there is a clear relationship between structure and property. If small structural differences lead to significantly different properties those methods will have difficulties to capture this.

If there is a lack of experimental data, or if there is no clear relationship between structure and activity, it is not recommended to build group contribution methods or chemoinformatics models. One could use molecular simulations for screening purposes. However, this requires accurate force fields parametrized with experimental data. It also takes significant time (several months/years) and resources and one should ask the question whether it is not preferable to use those for an experimental campaign.

The use of quantum mechanical modeling like DFT, where the solvent is not explicitly considered, is highly inaccurate, even random, and not recommended.

References

1. Kohl, A. and Nielsen, R. (1997), "Gas Purification," *Gulf Professional Publishing*, June, pp. 67-75.
2. Jablonka, K., Ongari, D., Moosavi, S. and Smit, B. (2020) "Big-data science in porous materials: materials genomics and machine learning", "*Chemical Reviews*", 120(16), pp 8066.

3. Varnek, A. and Bashkin, I. (2001) "Chemoinformatics as a theroretical chemistry discipline", "Molecular Informatics", 30, pp 20.
4. Yang, X., Rees, R. et al. (2017) "Computational modeling and simulation of CO_2 capture by aqueous amines", "Chemical Reviews", 117(14) pp 9524.
5. Marenich, A., Cramer, C. and Truhlar, D. (2009). "Universal solvation model based on solute electron density and on a continuum model of the solvent defined by the bulk dielectric constant and atomic surface tensions" "*The Journal of Physical Chemistry B*", 113(18), pp 6378.

604592 Comparative Studies of Energy Saving Performance for CO_2 Capture

Abstract:

Carbon dioxide (CO_2) capture, storage and utilization technology has attracted recent attention as a technology to reduce greenhouse gas emissions and mitigate global warming. On the contrary, CO_2 demands have been increased for inactive gases, dry ice in markets and refrigerant instead of Chlorofluorocarbons (CFCs). So far we have captured CO_2 from the source with high CO_2 concentration due to the limitation of CO_2 demands. However, we will have to capture CO_2 from the source with lower CO_2 concentration due to recent diversification of CO_2 demands as well. Therefore, it is necessary to design suitable CO_2 capture processes for satisfying to several CO2 demands and sources.

In this research, therefore, characteristics of several CO_2 capture processes have been investigated and the required energy for these processes have been compared from the process design point of view in order to create a tool for designing suitable processes by simulations and experiments.

600038 Design of a Fully Integrated Artificial Photosynthetic System for a Moisture-Gradient CO_2 Capture and Reduction

Abstract:

The effect of anthropogenic emissions of CO_2 on the climate change is becoming progressively prevalent. The current concentration of CO_2 in the air is 400 ppm, which is increasing at a rate of ~1.9 ppm per year. There exist several technologies such as amine scrubbing, oxyfuel combustion, and calcium looping to capture CO_2 from flue gases, which are inherently energy intensive. However, there are only a few techniques such as moisture-swing adsorption that can capture CO_2 from the air but requires modulation of humidity and downstream separation. The captured CO_2 can be either sequestered at the storage site or chemically transformed into a usable product. Integrating CO_2 capture with reduction not only eliminates the need for additional transportation and storage but also increases the efficiency of CO_2 capture. Here we provide a conceptual design of an integrated system where an anion-exchange resin for CO_2 capture is directly connected to a photo/electrochemical cell for CO_2 reduction. The CO_2 capture in this resin works on a moisture-gradient absorption of CO_2 accompanied by a carbonate/bicarbonate equilibrium. The dry-side of the resin captures CO_2 from the air and produces a high concentration of HCO_3^- diffusing towards the wet-side of the resin, where HCO_3^- converts back to CO_2 at the attached cathode to produce fuels. The limiting current density of an integrated system as a function of resin properties, catalysts activity and selectivity, electrolyte pH, physical dimensions and relative humidity are simulated using COMSOL Multiphysics. The limiting current density increases with increasing the resin conductivity, and mass-transfer coefficient of CO_2 in air. The efficiency of such integrated system using air as well as flue gas will be discussed as a part of this work.

608504 Study on CO Conversion from CO_2 By Chemical Looping Process

Abstract:

Currently, in order to reduce CO_2 emissions in the worid, research and development of more efficient power generation technologies and renewable energy technologies that do not emit CO_2 are being conducted. The research and development of high efficiency power generation technology conducts improving the performance of turbines used in thermal power plants, and the research and development of renewable energy technology conducts photovoltaic power generation, wind power generation, and biomass power generation. Particularly, research on renewable energy technology has been carried out actively, and the development of virtual power plants (VPP) for electricity exchange has been carried out. In the future, thermal power generation is expected to be replaced by renewable energy sources as the main source of electricity supply, and power generation plants using fossil fuel are expected to decrease. However, since renewable energy converts energy such as sunlight into electricity, the renewable energy itself has only value as electricity. On the other hand, fossil fuels are not only used as a fuel source for thermal power generation, but also in the manufacturing industry, with the steel industry as a typical destination. In the steel industry, fossil fuels are used as blast furnace fuel and in blast furnace as coke feedstock which is an iron ore reducing material. As can be seen from the fact that fossil fuels are coke feedstock, fossil fuel contains carbon atoms that can be converted from gasification reactions to syngas and converted into chemical products, and thus have value not only as a fuel but also as a raw material for chemical products. The demand for fossil fuels in the manufacturing industry will not disappear even if CO_2 emission free power generation is achieved through the advancement of renewable energy technologies. On the contrary, the demand for fossil fuels is expected to increase with the economic development of developing countries. As the global CO_2 emissions are expected to increase, not only renewable energy technologies to reduce CO_2 emissions in the power generation sector, but also carbon dioxide capture and storage (CCS) and carbon dioxide capture and utilization (CCU) technologies to reduce CO_2 emissions are attracting attention. CCS is a technology to separate and store high concentrations of CO_2 emitted from thermal power plants in the ground, but it will take more than 10 years to determine if the area is capable of storage, and it is estimated that it will be difficult to secure storage capacity. Therefore, CCU designed for effective utilization of CO_2 are more important than CCS that only store CO_2. CCU are currently being studied for biofuel production by microalgae and natural gas reforming, which can produce CO from fossil fuels. CO can be converted into not only a fuel but also a chemical product by the conversion process. CCU has received a lot of attention, but there are still challenges. Biofuel production by microalgae has not been put to practical use on a scale that can significantly reduce CO_2 emissions due to its slow CO_2 conversion rate. Although natural gas reforming is a thermochemical reforming process and therefore has a faster CO_2 conversion rate than biofuel production by microalgae, it is essential to combine natural gas reforming with unused waste heat, such as industrial waste heat, because of the need for an energy source. Water gas reverse shift reaction is one of the existing technologies for CO production using CO_2. Since water gas reverse shift reaction is an equilibrium reaction, the removal of compounds other than CO is necessary, and it remains a problem as a CO production process. Water gas reverse shift reaction chemical looping process (RWGS-CLP) has been developed while the CO_2 based CO production technology still has some problems. RWGS-CLP is a process to produce CO by transferring oxygen between H_2 and CO_2

using an oxygen carrier, instead of reacting H_2 and CO_2 at the same time, as in water gas reverse shift reaction. In RWGS-CLP, it is a process to produce CO by reduction the oxygen carrier with H_2 and then oxidization oxygen carrier with CO_2, and since the redox reaction is separated, CO can be produced at high concentrations with few impurities in the production gas. In previous studies, iron-based composite oxides containing La and Sr have been used as oxygen carriers, but La and Sr are not economical due to their high cost. In this presentation, RWGS-CLP was performed using iron-based composite oxides containing inexpensive Ca and Zn as oxygen carriers, and CO_2 was reduced to CO. The iron-based composite oxide containing Ca is called calcium ferrite and is used for chemical looping combustions, and it is reported to have high oxygen transfer rate. Although it has been reported that iron-based composite oxides containing Zn were capable of redox at 573 K, there is no report of their application in RWGS-CLP. In this presentation, the reaction behavior of these oxygen carriers in the redox reaction with H_2 and CO_2 and their physical properties after redox are evaluated, and the performance of the various oxygen carriers in RWGS-CLP is reported. In the experiments, thermogravimetric analysis and GC analysis using fixed bed reactor were carried out. In thermogravimetric analysis, the oxygen carrier was set in the thermogravimetric analysis device, and it was evaluated reaction rate under isothermal conditions. In GC analysis using fixed bed reactor, the redox reaction was carried out under isothermal conditions, and the gas composition of the production gas was analyzed by GC.

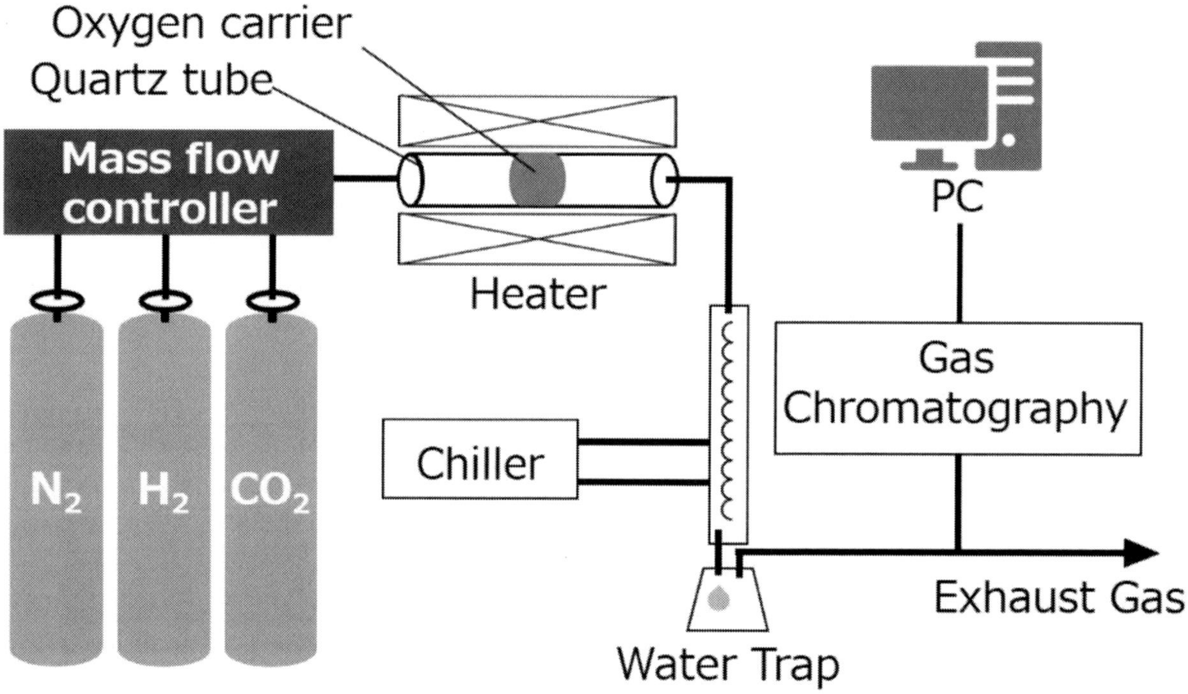

Figure. Schematic diagram of Chemical Looping Process

609258 CO_2 Solubility in Water Confined By Kerogen Nanopore in Relation to CO_2 Sequestration in Depleted Shale Reservoirs: A Molecular Dynamics Simulation Study

Abstract:

Depleted shale reservoirs are feasible geological media for CO_2 sequestration. CO_2 solubility trapping plays an important role because massive water presents in the depleted shale reservoirs due to connate water, hydraulic fracturing, and water invasion. Besides, kerogen is a major component of shale, generating an extensive number of porous media at nanometer scale. The properties of kerogen (e.g., wettability) are significantly changed with different thermal maturities. In this work, we use graphene oxide (GO) to model kerogen. Various oxidized degrees (ODs) from 0.04 to 0.16 are used to represent different maturities of kerogen. Three pore sizes (1, 2, and 4 nm) and two hydroxyl patterns (regular and random) are designed to describe different kerogen nanopores. The results indicate that, as OD increases (kerogen maturity degree decreases), CO_2 adsorption amount dramatically decreases, and CO_2 solubility also decreases. CO_2 solubility is generally enhanced compared to bulk solubility in the kerogen nanopores except under the super low maturity kerogen and super small nanopore ($\leq$1 nm). As pore size increases, CO_2 solubility increases when the kerogen is at low maturity, while it decreases when the kerogen is at high maturity. The random hydroxyls pattern cases always have higher CO_2 solubility than those regular pattern cases and the difference of water alignment in these two pattern cases can be attributed to the hydrogen bonds structure between water and hydroxyls. From CO_2 sequestration perspectives, the high maturity kerogen nanopore is more capable to store CO_2. Besides, in the high maturity kerogen nanopores, small nanopores can sequestrate more CO_2 than large nanopores under the equivalent pore volume, and it reverses in the low maturity kerogen nanopores.

CHARACTERIZATION OF FACTORS AFFECTING THE GAS WELL DELIVERABILITY OF THE JIAOSHIBA SHALE GAS FIELD BY MULTIVARIATE STATISTICAL ANALYSIS

Yizhong Zhang[1,2], Maolin Zhang[1], Bin Ju[3], Jincai Shen[4], Lan Ge[4], Fanhua Zeng[2]*

**Corresponding author. Email address: Fanhua.Zeng@uregina.ca.*

[1] Hubei Cooperative Innovation Center of Unconventional Oil and Gas, Yangtze University, Hubei, Wuhan, China, 430100

[2] University of Regina, Regina, SK, Canada, S4S 0A2

[3] School of Oil and Natural Gas Engineering, Southwest Petroleum University, Chengdu, Sichuan, China, 610500

[4] SINOPEC Chongqing Fuling Shale Gas Exploration and Development Co., Ltd., Chongqing, China, 408014

Abstract

The Longmaxi shale gas in the Jiaoshiba Shale Gas Field is the first and the largest successfully developed shale gas in China. The analysis of key factors of deliverability of gas wells plays a significant role in efficiently developing shale gas reservoirs. However, there are few methods analyzing the key factors of deliverability of shale gas wells by considering both geological characters and engineering parameters. In this study, the reservoir characteristics, drilling and completion data, fracturing data (including the well logging interpretation of horizontal section, the track and layer section that the well bore passed through, and fracturing parameters) and production data of 86 gas wells in the Wufeng-Longmaxi shale, Jiaoshiba Shale Gas Field in Eastern Sichuan Basin, China are statistically analyzed and ordered by multivariate statistical analysis. The gas well deliverability is described by absolute open flow potential and 9 factors (physical parameters of reservoir, including density of rock, porosity, total organic carbon, brittle mineral content of rock and drilling and fracturing engineering quality, including horizontal length of gas well, horizontal length ratio through different stages in the formation, number of fracturing stages, liquid volume, and sand amount) that may affect the deliverability of shale gas wells are considered. The results reveal that the absolute open flow potential has a close relationship with porosity, total organic carbon, brittle mineral content and total liquid volume. The multivariable nonlinear regression models of absolute open flow potential with different key factors considered are built, and the prediction results are reliable by goodness-of-fit test acceptable with real gas well deliverability testing data. The research methods and results in this study can provide a reference basis for reasonable production rate allocation, design, and adjustment of development plan for the shale gas reservoirs.

Keywords: Gas well deliverability, Longmaxi shale gas, Jiaoshiba Shale Gas Field, Multivariate statistical analysis, Multivariable nonlinear regression

1. Introduction

Following the increased demand of petroleum energy and the developed science and technology, the industry nowadays has focused on the unconventional resources. Shale gas reservoirs are one of the most popular kind of unconventional resources with extremely low permeability, tight porous structure, large specific area and high heterogeneity, which have attracted attentions of researchers worldwide [1-3]. Shale gas has self-generated, self-contained, organic-rich, nanoscale pore structure, and other unusual characteristics and it could be considered as a source as well as a reservoir [4-5]. In addition to free gas and small amount of dissolved gas, adsorbed gas is largely existed in the tight formation, which makes it difficult to produce owing to the complexity of the reservoir [6-8]. The Sichuan-Chongqing region in southwest China is one of the fastest shale gas production growing regions in the world. The annual shale gas deliverability of China has reached 2×10^{10} m^3 with the annual shale gas production of 1.5×10^{10} m^3 [9]. As one of the main contributors of natural gas production growth in China, the gas production of Jiaoshiba shale gas reservoir has exceeded the daily production of 1.67×10^7 m^3/d since 2013 [10]. The large production of commercial shale gas flow of the Longmaxi Formation (e.g. more than 2×10^5 m^3/d of JY1 well) has made the Jiaoshiba area one of the most important shale gas field in China.

Compared with typical shale gas reservoirs in North America, the shale gas in Sichuan Basin has lower TOC value, lower porosity, lower gas saturation, smaller thickness and higher buried depth, which makes its exploration more complicated [9]. In both the early test stage and current production stage, there are large differences among deliverability of the shale gas wells in the Jiaoshiba area. Therefore, the deliverability evaluation is the basis of reasonable shale gas production and prediction, and the research on deliverability evaluation and its related reservoir and engineering parameters of shale gas wells is particularly important. Previous studies have focused on the impacts of reservoir quality including sedimentary environment [11], structure [12-13], mineralogy and lithology [14-17] and engineering quality including drilling and completion techniques [12, 18] and multi-stage hydraulic fracturing performance [19-20] on the shale gas production. Their results revealed that both the reservoir quality and engineering quality will affect the produced gas rate of the shale gas wells. Most of researches did analysis based on the relationship between the interested variable and its single impacting factor, comprehensive multivariate statistical analysis with consideration of correlation and collinearity among factors needs to be further studied. Moreover, instead of the produced gas rate or gas bearing content, the absolute open flow (AOF) potential may be a better parameter to describe the deliverability of gas well. The AOF potential is a hypothetical gas flow rate that may be produced if the flowing pressure at the bottom of the well (BHFP) were equivalent to the atmospheric pressure, i.e. the maximum flow rate that the gas well may produce [21]. In this study, the well deliverability of Jiaoshiba shale gas is described by the AOF potential based on BHFP and the impacts of reservoir and engineering characteristics on the AOF potential will be comprehensively analyzed. The data of over 86 wells of Fuling Jiaoshiba shale gas field is analyzed and the impacts of the reservoir quality (i.e. density of rock, porosity, total organic carbon, brittle mineral content of rock) and the engineering quality (i.e. horizontal length of gas well, horizontal length ratio through different formation stages, number of fracturing stages, liquid volume, and sand amount) on the AOF are quantified. The most important factors affecting deliverability of the gas wells in the Jiaoshiba area will be obtained by hierarchical clustering analysis method. The regression correlation between the well deliverability and the important parameters will be built. The objective of this study is to find a new method to evaluate the key factors that affect the well deliverability of shale gas reservoir by multivariate statistical analysis with consideration of correlation and collinearity among factors and build multivariable nonlinear regression models based on these key factors, so that the well deliverability of shale gas reservoir could be predicted.

2. Shale gas reservoir characters

The Longmaxi shale gas in the Jiaoshiba area located in the Eastern Sichuan Basin in the southwest region of China is the first and the largest successfully developed shale gas reservoir in China [22]. The shale gas bearing formation is the Upper Ordovician Wufeng Formation to Lower Silurian Longmaxi Formation with a thickness of about 200–300 m, area of 272.44 km^2 and buried depth of less than 3500m. Comprehensively considering the lithology, electrical and physical properties, geochemical and gas-bearing characteristics, the gas bearing shale section in the Upper Ordovician Wufeng Formation to Lower Silurian Longmaxi Formation could be vertically divided into three parts with stable horizontal stratigraphic distributions for each member (i.e. Long1, Long2 and Long3 from bottom to top). Long1 member is rich in fossils (e.g., radiolarian, calthrop, and graptolite) and is composed of black mudstone and argillaceous siltstone. The Long2 member consists of argillaceous siltstone and fine-grained sandstone. The Long3 member is composed of dark-gray mudstone and gray shale with fewer fossils [5, 23].

Previous studies suggested that Long1 member consisted of one 3-ord sequence, and could be divided into three system tracts (i.e. the transgressive system tract (TST), early highstand system tract (EHST), and late highstand system tract (LHST)) [10, 15, 16]. As shown in **Fig.1** [5], this study is targeted in the Long1 member, which is abundant of organic matter with pores and could be divided into 3 stages (i.e. stage1, stage2 and stage3 from bottom to top) based on physical properties with gas bearing increasing from top to bottom. The Long1 member is of good brittleness, and the brittle mineral content gradually increases from top to bottom. The natural gas in the layer belongs to over-mature dry gas, in which methane is the main component without containing hydrogen sulfide, carbon monoxide and hydrogen.

3. Samples

The AOF potential is a hypothetical gas flow rate that may be produced if the flowing pressure at the bottom of the well (BHFP) were equivalent to the atmospheric pressure [21]. The AOF potential may be a better parameter to describe the deliverability of the gas wells as it gives the maximum flow rate that the gas well can produce. As shown in Fig.2, the AOF potential data of and 9 factors that may affect the deliverability of gas wells (i.e. horizontal length, number of fracture stages, total liquid volume, total amount of sand, rock density, porosity, TOC, brittle mineral content of rocks, horizontal length ratio through stage 1 and stage 3 of Long1 member) are collected (data is offered by SINOPEC Chongqing Fuling Shale Gas Exploration and Development Co., Ltd., Chongqing, China). Though **Fig.2** shows certain relationship between AOF potential and single impacting factor, the correlation and collinearity among different factors are not considered. According to the data, the average reservoir pressure of the Long1 member is 37.69MPa with pressure coefficient 1.55. The reservoir temperature is 85.99°C. The pay zone thickness is between 20m-200m and the most favorable thickness is about 38m. The kerogen type of the organic matter is mostly type II_1 to II_2. The total organic carbon (TOC) contents of the shale range from 1%-8% with average value of 5% and it is increasing from top to bottom. The vitrinite reflectance (R_o %) of organic matter ranges from 2.5%-3.7% with average value of 2.9%. The porosity of the shale is between 1.88% to 7.06%, with average value of 4.15% and the permeability of the shale is less than 0.001mD. The horizontal wells in the sample have lateral length between 1500m to 1800m, well spacing between 500m to 600m, and the distance between adjacent clusters between 20m to 25m. All the horizontal wells are multi-stage hydraulic fractured with number of fracturing stages between 15 to 29. The fluid volume per stage is between 1500m^3 to 1800m^3, and proppant volume per stage is between 50m^3-65m^3. The fractured half-lengths range from 300m to 350m, and propped half-lengths range from 250m to 270m. The fracture heights range from 40m to 50m, and main fracture width range from 5mm to 10mm.

4. Methods

4.1. Normality test

When working with multivariable regression problems, some assumptions should be followed. The regression process of independent variables and the target variable are assumed to follow Gauss–Markov theorem (i.e. the ordinary least squares (OLS) estimator has the lowest sampling variance within the class of linear unbiased estimators, if the independent variables have no perfect collinearity, and the errors in the regression model are uncorrelated with equal variances and expectation value of zero) [24]. Thus, the first step of the multivariate statistical analysis is to test whether the sample variables follow normal distribution. The Q–Q (quantile-quantile) plot is applied, which is a probability plot that shows a graphical comparison between the samples and normal distribution by plotting their quantiles against each other [25]. If the points in the Q–Q plot approximately lie on the 45-degree line (i.e. $y = x$), the distribution of the sample being compared are similar enough to normal distribution. If normality test of a variable behaves not well, certain transformation (i.e. logistic transformation, logarithm transformation, index transformation, power transformation, etc.) of the variable and nonlinear regression model should be considered.

4.2. Standardization

Before the key factor analysis, the calculation of correlations between each parameter is essential. Since the units, scales and ranges of the factors are different, the first step is to standardize all the parameters based on the normality test results. The method applied in the study is the Fisher Z-transformation method. The standard score (Z score) for each data point is its difference from sample mean divided by the sample standard deviation.

$$Z = \frac{X - \overline{X}}{s} \tag{1}$$

where

X is the data point.
$\overline{X}$ is the sample mean.
s is the sample standard deviation.

4.3. Correlation analysis

After standardization, the correlations between each factor are calculated. This is to eliminate the repeated consideration of correlation and collinearity in the later multivariable regression. In other words, if two variables are strongly related, only the principle variable between them is considered in later analysis. The sample data has a number of n wells, and each well has a number of k variables including a dependent variable (X_k) which is Z scores for AOF potential and $k-1$ independent variables (X_1, X_2, ..., X_{k-1}) which are Z scores for the factors affecting AOF potential. The sample data could be shown in matrix form as follows

$$X = \begin{bmatrix} x_{1,1} & x_{1,2} & \cdots & x_{1,k} \\ x_{2,1} & x_{2,2} & \cdots & x_{2,k} \\ \cdots & \cdots & & \cdots \\ x_{n,1} & x_{n,2} & \cdots & x_{n,k} \end{bmatrix} \quad (2)$$

Then the correlation matrix for all variables would be

$$R = \begin{bmatrix} r_{11} & r_{12} & \cdots & r_{1k} \\ r_{21} & r_{22} & \cdots & r_{2k} \\ \cdots & \cdots & \cdots & \cdots \\ r_{k1} & r_{k2} & \cdots & r_{kk} \end{bmatrix} \quad (3)$$

$r_{ij}(i,j=1,2,...,k)$ is the correlation coefficient for $x_{ij}(i=1,2,...,k)$ and $x_{ij}(i=1,2,...,k)$, and the calculation equation is

$$r_{ij} = \frac{\sum_{m=1}^{k}(x_{im}-\overline{x_i})(x_{jm}-\overline{x_j})}{\sqrt{\sum_{m=1}^{k}(x_{im}-\overline{x_i})^2\sum_{l=1}^{k}(x_{jl}-\overline{x_j})^2}} \quad (4)$$

The value of correlation coefficient should be between -1 and 1. If the correlation coefficient is close to 1, the involving two variables have a strong positive relationship, and if it is close to -1, the involving two variables have a strong negative relationship. If the correlation coefficient is close to 0, this suggests that the involving two variables may be independent to each other.

4.4. Hierarchical clustering analysis

As long as correlations among factors are obtained, clustering method could be applied to find the key factors while eliminate the factors with small effect on AOF potential or strong correlation with other factors. Hierarchical clustering analysis (HCA) is a method to find the hierarchy relationship among variables, so that the variables could be divided into clusters [26]. HCA is performed on the normalized sample data set of the 86 well data including AOF potentials and 9 different factors, using centroid linkage clustering (Eq.5) as linkage criteria. The method will try to link the two most related factors in the data set to form a new cluster at each step (**Fig.3** agglomerative hierarchical clustering). The new merged cluster is then considered as a variable in the relation map and agglomerative approach is again used. Finally, a tree-hierarchy structure with linkage lines that join all the factors would be built.

$$D_{uv}^2 = (\bar{X}_u - \bar{X}_v)'(\bar{X}_u - \bar{X}_v) \tag{5}$$

, where $\bar{X}_u$ is the centroid for cluster C_u. $\bar{X}_v$ is the centroid for cluster C_v.

4.5. Multivariable nonlinear regression

Once the main factors that affect the AOF potential are found, other factors are either unessential to AOF potential or have strong correlation to the main factors, which could be eliminated. The next step is to linearize the main factors, including logistic transformation, logarithm transformation, index transformation, power transformation, etc. Then, a set of multivariable linear regressions is carried out to find the optimal regression model. If there are k variables, the linear regression for the variables is

$$y_i = \alpha + \beta_1 X_{i1} + \ldots + \beta_k X_{ik} + \varepsilon_i = \alpha + \sum_{j=1}^{k} \beta_j X_{ij} + \varepsilon_i, i = 1,2,\ldots,n \tag{6}$$

Least squares could be used to find the intercept a and coefficients β_j (j=1,2,...,k), which is to find the optimums for a and β_j (j=1,2,...,k) when the sum of squared residuals, *SSe*, is minimized. The squared residual r_i^2 is defined as the difference between real value and predicted value of the dependent variable.

$$r_i = y_i - \hat{y}_i = y_i - f(x_i, \alpha, \beta) \tag{7}$$

$$SSe = \sum_{i=1}^{n} r_i^2 \tag{8}$$

$\hat{y}_i$ is the estimation value of dependent variable y_i. $a, b_j (j = 1,2,\ldots,k)$ are the estimation values of intercept and coefficients $\alpha, \beta_j (j = 1,2,\ldots,k)$. Let the partial derivatives with respect to $a, b_j (j = 1,2,\ldots,k)$ equals 0.

$$\begin{cases} \dfrac{\partial SS_e}{\partial a} = 0 \\ \dfrac{\partial SS_e}{\partial b_j} = 0 \end{cases} \quad j = 1,2,\ldots,k \tag{9}$$

It could be also proved that $a, b_j (j = 1,2,\ldots,k)$ is the unbiased estimation of $\alpha, \beta_j (j = 1,2,\ldots,k)$.

5. Results and Discussions

5.1. Normality test

Fig.4 shows the Q–Q (quantile-quantile) plots for the 10 variables of the 86 wells. The results reveal that the points of most variables in the Q–Q plots approximately lie on the 45-degree line, which means they follow normal distribution well. The only variable may not behave with good normality is horizontal length ratio through stage 1 and stage 3 of Long1 member. This variable may need transformation (i.e. logistic transformation, log transformation, logistic transformation, etc.) in the later regression study. Thus, the multivariable nonlinear regression would be needed in this study.

5.2. Standardization and correlation of the variables

The matrix of correlation coefficients of all variables is shown in **Table 1**. **Table 1** suggests that there is no strong linear relation between the AOF potential and single one factor. Relatively, the AOF potential is positively related to the TOC, the brittle mineral content, and the total liquid volume. It is negatively related to rock density. Some factors show strong correlation with each other, and the number of fracture stages and total liquid volume have the highest correlation coefficient of 0.946. The drilling and fracturing engineering factors, including the number of fracturing stages, the length of horizontal well, the total liquid volume and the total sand amount, are highly linear related. Among the geological factors, the rock density is highly negative to TOC, brittle mineral content, and the TOC and the brittle mineral content of rock are highly linear related. The collinearity among some factors is very strong and may affect AOF potential modelling, which should be eliminated in the multivariate statistical analysis of AOF potential.

5.3. Hierarchical clustering analysis

According to **Table 1**, the number of fracturing stages and total liquid volume have the highest correlation coefficient, which is 0.946. These two factors could be firstly clustered together, and the central point of this cluster is calculated by centroid method. Then the correlation analysis is done again after applying the whole cluster as a variable. This progress is continued until all the variables converged to one cluster. The result of hierarchical clustering analysis with linkage lines is shown **Fig.5**, and **Table 2** gives the importance order of the effect of each factor on the AOF potential.

It is suggested in the hierarchical clustering analysis results that the order of the factors that affect AOF potential is: a. porosity, b. TOC, c. brittle mineral content of rock, d. horizontal length ratio through stage 1 and stage 3 of Long1 member, e. number of fracturing stages, f. total liquid volume, g. horizontal length, h. total sand amount and i. rock density. The AOF potential firstly has closest relationship to porosity. It then has closer relationship to the cluster of TOC, brittle mineral content of rock, horizontal length ratio through stage 1 and stage 3 of Long1 member. These four factors belong to the first cluster. The other four factors including number of fracturing stages, total liquid volume, horizontal length and total sand amount belong to the second cluster. Rock density belongs to the third cluster. Thus, the main factors that affect AOF potential are porosity, TOC, brittle mineral content of rock and number of fracturing stages.

5.4. Multivariable nonlinear regression

After the main clusters in the hierarchical clustering analysis are obtained, the key factors that affect the AOF potential are found. The independent variables in the multivariable nonlinear regression models could be chosen according to the cluster results. With consideration of the correlation and collinearity among factors, one factor in each cluster with smallest linkage distance is chosen in the regression model, and four models after testing (**Table 3**) are regressed in this study.

With trials of nonlinear transformation of the involving variables, the linear regression of the transformation is ideal, and all the coefficients have passed the student t test for significance. The four models are shown as follows.

Model 1:

$$Y' = \ln Y, X_1' = \ln X_1, X_2' = (\ln X_2)^8, X_3' = X_3 \tag{10}$$

$$Y' = -13.067 + 1.509X_1' + 0.012X_2' + 0.867X_3' \tag{11}$$

Model 2:

$$Y' = \ln Y, X_4' = \ln X_4, X_2' = (\ln X_2)^8, X_3' = \ln X_3 \tag{12}$$

$$Y' = -1.756 + 1.465X_4' + 0.011X_2' + 1.043X_2' \tag{13}$$

Model 3:

$$Y' = \ln Y, X_4' = \ln X_4, X_5' = X_5^{\frac{1}{100}}, X_2' = (\ln X_2)^8, X_3' = \ln X_3 \tag{14}$$

$$Y' = -1.603 + 1.434X_4' + 0.226X_5' + 0.011X_2' + 0.848X_3' \tag{15}$$

Model 4:

$$Y' = \ln Y, X_1' = \ln X_1^{\frac{1}{10}}, X_3' = X_3^{\frac{1}{100}}, X_2' = (\ln X_2)^8, X_6' = X_6 \tag{16}$$

$$Y' = -12.556 + 5.211X_1' + 0.149X_5' + 0.012X_2' + 0.035X_6 \tag{17}$$

It is shown from **Fig.6** that the total AOF potential of 86 wells is 5420.48×$10^4$$m^3$/d and the total AOF potential of 86 wells predicted by the 4 models range from 5110.24×$10^4$$m^3$/d to 5607.74×$10^4$$m^3$/d. The average AOF potential of single well is 63.03×$10^4$$m^3$/d and the average AOF potential of single well predicted by the 4 models range from 59.42×$10^4$$m^3$/d to 65.21×$10^4$$m^3$/d. The coefficient of determination (R^2) in the figure is the proportion of the variance in the dependent variable Y that is predictable from the independent variables $X_j(j=1,2,...,k)$. The R^2 values show that all the 4 models could be relatively preferable models to predict the AOF potentials based on different factors. In total, the prediction value of model 1 and model 4 are closer to the real AOF potential value with higher coefficients of determination (R^2) and less prediction errors.

6. Conclusions

In this study, instead of the produced gas rate, the absolute open flow (AOF) potential is analyzed as a better parameter to describe the deliverability of gas well. The reservoir parameters, drilling and completion data, fracturing data (including the well logging interpretation of horizontal section, the track and layer section that the well bore passed through, fracturing scale and monitoring parameters) and sample production data of 86 gas wells in the Wufeng-Longmaxi shale, Jiaoshiba shale gas field in Eastern Sichuan Basin, China are considered in the multivariate statistical analysis. The hierarchical clustering analysis could give the orders of the key factors that affect gas well AOF potential as well as eliminate the collinearity among the factors, which gives the basis of multivariable nonlinear regression. The results suggest that the order of the factors affects AOF potential is: a. porosity, b. TOC, c. brittle mineral content of rock, d. horizontal length ratio through stage 1 and stage 3 of Long1 member, e. number of fracturing stages, f. total liquid volume, g. horizontal length, h. total sand amount and i. rock density. The estimation results of the AOF potential by the multivariable nonlinear regression models based on the key factors from hierarchical clustering analysis have small differences to that of the actual values of the sample with goodness-of-fit test acceptable. The predictions of AOF potential of the models are reliable, and the feasibility of this multivariate statistical analysis is verified in the study of shale gas well deliverability. The research results can provide a reference basis for design of drilling and completion, the reasonable production rate allocation, design and adjustment of development plan of the shale gas reservoirs.

Acknowledgement

The authors would like to acknowledge the support from National Natural Science Foundation of China (NSFC Grant Number: 520004032) and the Thirteen-Five National Science and Technology Major Projects of China: 1. Study on Shale Gas Fracturing and Gas Production Regularity in Wufeng Formation and Longmaxi Formation in Fuling Area (Sub-Issue Number: 2016ZX05060-026) and 2. Research on the efficient development technology and policy of inter-salt shale oil reservoir (Sub-Issue Number: 2016ZX05049005-010).

References

1. Zolfaghari, A., Dehghanpour, H., & Xu, M. (2017b). Water sorption behaviour of gas shales: II. Pore size distribution. *International Journal of Coal Geology, 179*, 187-195.
2. Zhang, Y., Yao, S., Zhang, M., Zhou, X., Mei, H., & Zeng, F. (2018). Prediction of adsorption isotherms of multicomponent gas mixtures in tight porous media by the oil–gas-adsorption three-phase vacancy solution model. Energy & Fuels, 32(12), 12166-12173.
3. Zhang, Y., Zhang, M., Mei, H., & Zeng, F. (2019). Study on salt precipitation induced by formation brine flow and its effect on a high-salinity tight gas reservoir. *Journal of Petroleum Science and Engineering, 183*, 106384.
4. Dong, T., Harris, N. B., Ayranci, K., Twemlow, C. E., & Nassichuk, B. R. (2017). The impact of composition on pore throat size and permeability in high maturity shales: Middle and Upper Devonian Horn River Group, northeastern British Columbia, Canada. *Marine and Petroleum Geology*, 81, 220-236.
5. Wang, C., Zhang, B., Hu, Q., Shu, Z., Sun, M., & Bao, H. (2019). Laminae characteristics and influence on shale gas reservoir quality of lower Silurian Longmaxi Formation in the Jiaoshiba area of the Sichuan Basin, China. *Marine and Petroleum Geology, 109*, 839-851.
6. Cao, T. T., Song, Z. G., Wang, S. B., & Jia, X. (2015). A comparative study of the specific surface area and pore structure of different shales and their kerogens. *Science China Earth Sciences, 58*(4), 510-522.
7. Chalmers, G. R. L., & Bustin, R. M. (2007). The organic matter distribution and methane capacity of the lower cretaceous strata of northeastern British Columbia, Canada. *International Journal of Coal Geology, 70*(1-3), 223-239.
8. Chalmers, G. R. L., & Bustin, R. M. (2008). Lower cretaceous gas shales in northeastern British Columbia, Part II: Evaluation of regional potential gas resources. *Bulletin of*

Canadian Petroleum Geology, 56(1), 22-61.
9. Ma, X., Li, X., Liang, F. , Wan, Y. , & Guo, W. . (2020). Dominating factors on well productivity and development strategies optimization in Weiyuan shale gas play, Sichuan Basin, SW China. *Petroleum Exploration and Development, 47*(3), 594-602.
10. Shu, Y., Lu, Y., Chen, L., Wang, C., & Zhang, B. (2020). Factors influencing shale gas accumulation in the lower Silurian Longmaxi formation between the north and South Jiaoshiba area, Southeast Sichuan Basin, China. *Marine and Petroleum Geology, 111*, 905-917.
11. Wang, Y. , Dong, D. , Li, X. , Huang, J. , Wang, S. , & Wu, W. . (2015). Stratigraphic sequence and sedimentary characteristics of Lower Silurian Longmaxi Formation in the Sichuan Basin and its peripheral areas. Natural Gas Industry B, 2(2–3), 222-232.
12. Hu, M., Huang, W., & Li, J. (2018). Effects of structural characteristics on the productivity of shale gas wells: A case study on the Jiaoshiba Block in the Fuling shale gasfield, Sichuan Basin. *Natural Gas Industry B, 5*(2), 139-147.
13. Zhang, H., Wang, J., & Zhang, H. (2016). Investigation of the main factors during shale-gas production using grey relational analysis. *The Open Petroleum Engineering Journal, 9*(1).
14. Wei, C., Wang, H., Sun, S., Xiao, Y., Zhu, Y., & Qin, G. (2012, January). Potential investigation of shale gas reservoirs, Southern China. In *SPE Canadian Unconventional Resources Conference*. Society of Petroleum Engineers.
15. Wang, C., Zhang, B., Lu, Y., Shu, Z., Lu, Y., Bao, H., ... & Chen, L. (2018). Lithofacies distribution characteristics and its controlling factors of shale in Wufeng Formation-Member 1 of Longmaxi Formation in the Jiaoshiba area. *Petroleum Research, 3*(4), 306-319.
16. Wang, C., Zhang, B. Q., Lu, Y. C., Shu, Z. G., Lu, Y. Q., Bao, H. Y., ... & Chen, L. (2018). Lithofacies distribution characteristics and main development controlling factors of shale in Wufeng Formation-Member 1 of Longmaxi Formation in Jiaoshiba area. *Acta Pet. Sin., 39*(6), 631-644.
17. Kim, G., Lee, H., Chen, Z., Athichanagorn, S., & Shin, H. (2019). Effect of reservoir characteristics on the productivity and production forecasting of the Montney shale gas in Canada. *Journal of Petroleum Science and Engineering, 182*, 106276.
18. Curtis, M. E., Cardott, B. J., Sondergeld, C. H., & Rai, C. S. (2012). Development of organic porosity in the Woodford Shale with increasing thermal maturity. *International Journal of Coal Geology, 103*, 26-31.
19. Cipolla CL, Lolon EP, Erdle JC, Rubin B. Reservoir modeling in shale-gas reservoirs. *SPE reservoir evaluation & engineering*. 2010 Aug 1;13(04):638-53.
20. Zhu W, Ma D, Qi Q, Song Z, Yue M. Multi-zone coupling productivity of horizontal well fracturing with complex fracture networks in shale gas reservoirs. Natural Gas Industry B. 2018 Feb 1;5(1):80-9.
21. Mohammed, K. R., Stigliano, H., Vicente, M. G., & Angulo, R. J. (2010, January). Comparison of Different Deliverability Testing Techniques in Low Permeability Gas Reservoir and Its Impact on the Absolute Open Flow (AOF) Estimation. In *SPE EUROPEC/EAGE Annual Conference and Exhibition*. Society of Petroleum Engineers.
22. Guo, X., Qin, Z., Yang, R., Dong, T., He, S., Hao, F., ... & Liu, K. (2019). Comparison of

pore systems of clay-rich and silica-rich gas shales in the lower Silurian Longmaxi formation from the Jiaoshiba area in the eastern Sichuan Basin, China. *Marine and Petroleum Geology, 101*, 265-280.

23. Yang, R., He, S., Hu, Q., Hu, D., Zhang, S., & Yi, J. (2016). Pore characterization and methane sorption capacity of over-mature organic-rich Wufeng and Longmaxi shales in the southeast Sichuan Basin, China. Marine and Petroleum Geology, 77, 247-261.
24. Henderson, C. R. . (1975). Best linear unbiased estimation and prediction under a selection model. *Biometrics, 31*(2), 423-447.
25. Gnanadesikan, M. B. W. . (1968). Probability plotting methods for the analysis for the analysis of data. *Biometrika, 55*(1), 1-17.
26. Rokach, Lior, and Oded Maimon. (2005). "Clustering methods." Data mining and knowledge discovery handbook. Springer US, 321-352.

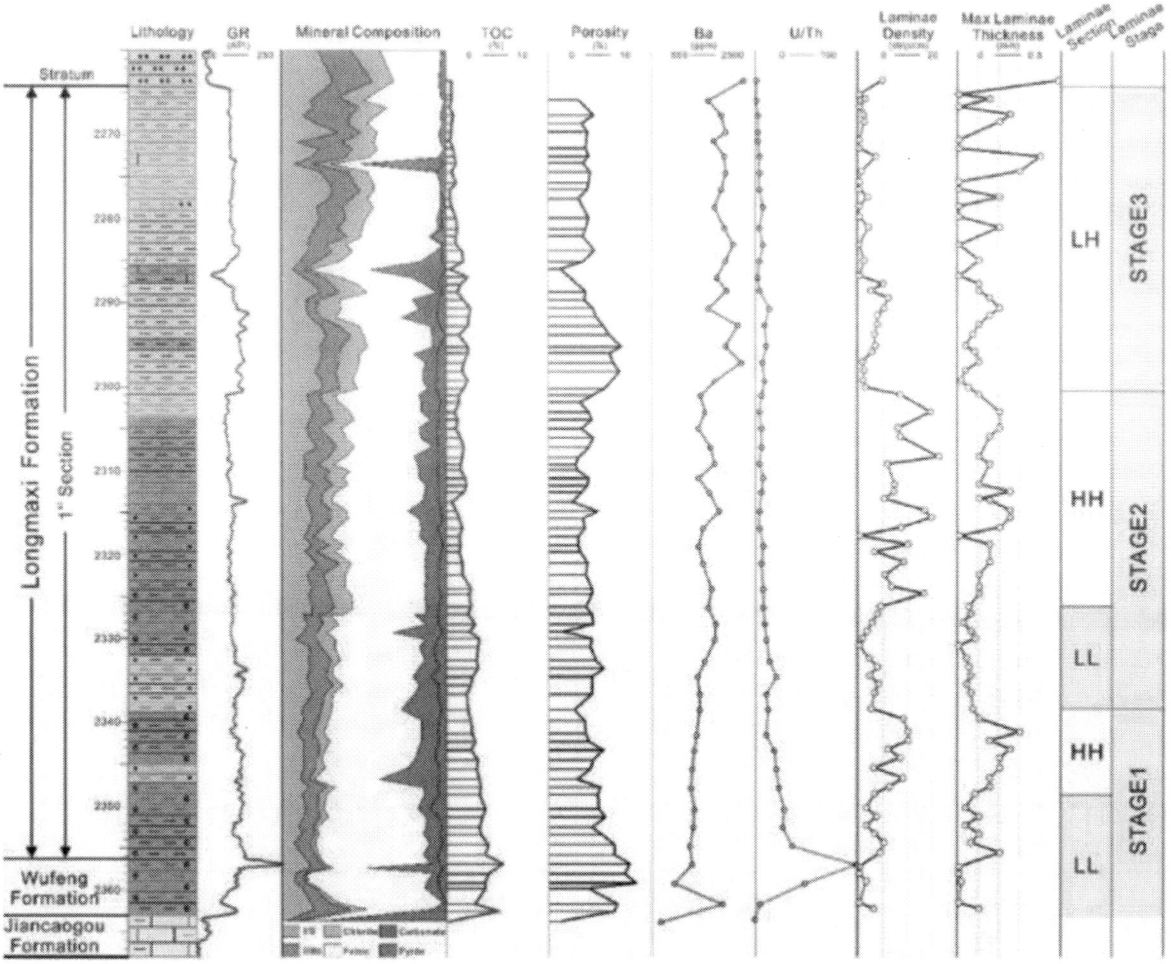

Fig.1 Lithofacies, gamma ray log, mineral content, TOC content, porosity, geochemical proxies, laminae density, and laminae stages of the Long1 member in well A (Wang et al., 2019).

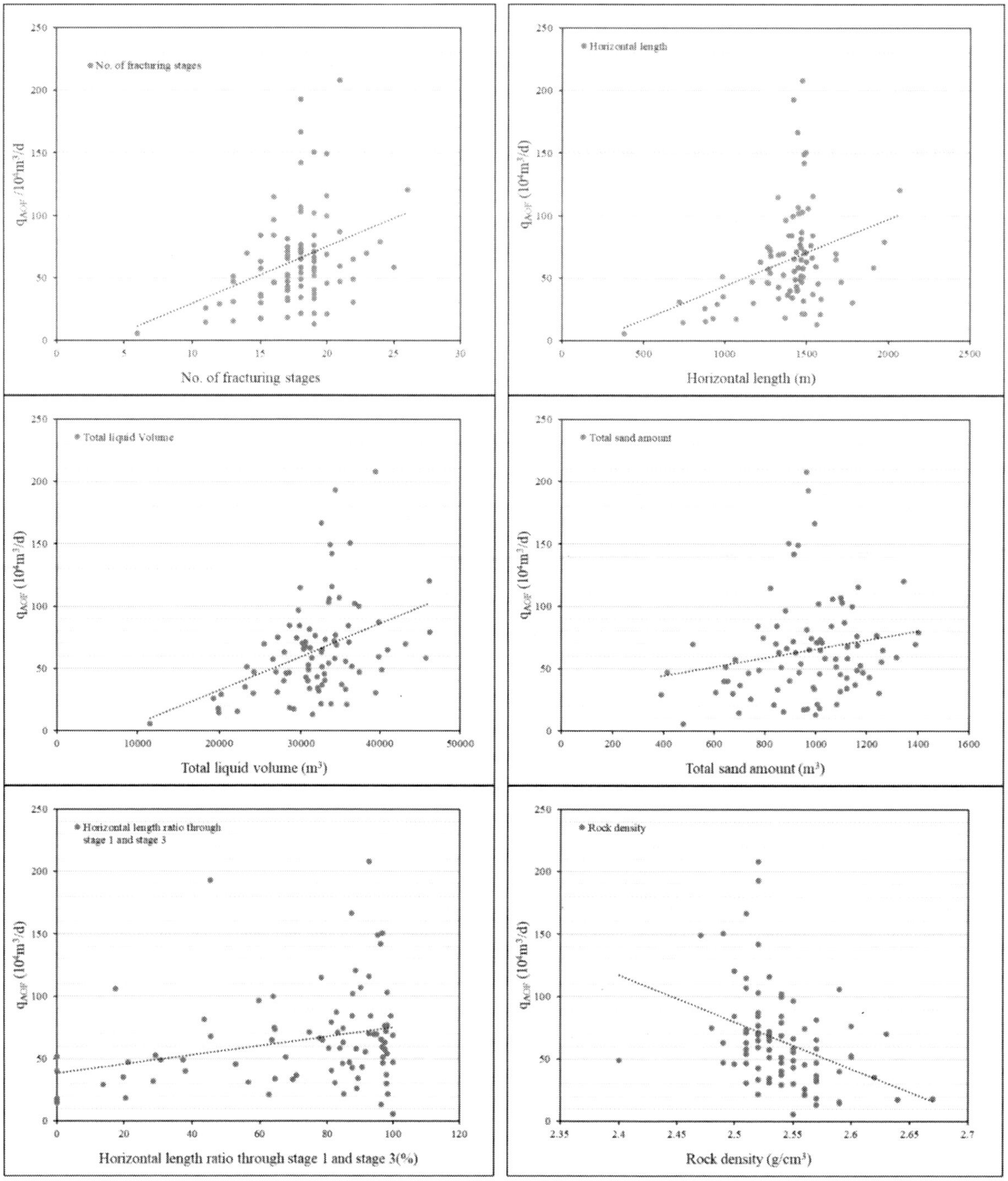
No. of fracturing stages
q_AOF (10^4m^3/d)
No. of fracturing stages
Horizontal length
q_AOF (10^4m^3/d)
Horizontal length (m)
Total liquid Volume
q_AOF (10^4m^3/d)
Total liquid volume (m^3)
Total sand amount
q_AOF (10^4m^3/d)
Total sand amount (m^3)
Horizontal length ratio through stage 1 and stage 3
q_AOF (10^4m^3/d)
Horizontal length ratio through stage 1 and stage 3(%)
Rock density
q_AOF (10^4m^3/d)
Rock density (g/cm^3)

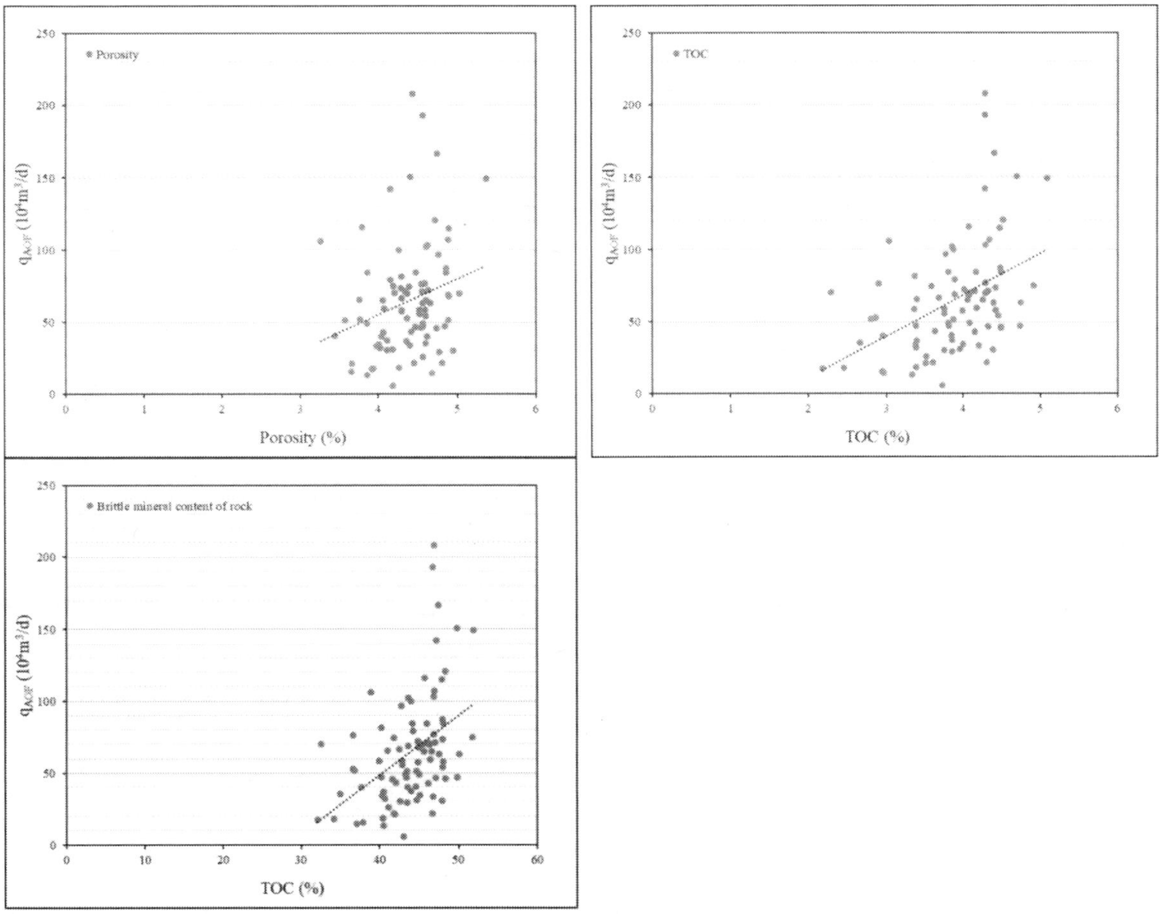

Fig.2 The scatter plots with single linear regression line of sample data of 86 gas wells in Jiaoshiba area with the AOF potential against 9 factors that may affect the deliverability of gas wells (i.e. number of fracture stages, horizontal length of well, total liquid volume, total sand amount, rock density, porosity, TOC, brittle mineral content of rocks, horizontal length ratio through stage 1 and stage 3 of Long1 member)

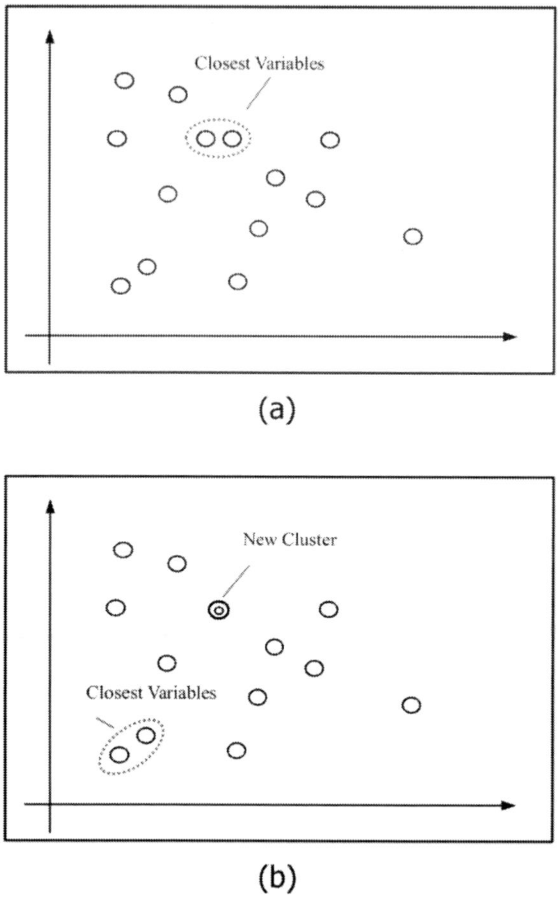

Fig.3 Hierarchical Clustering Analysis: (a) Clustering the variables by closest relation; (b) The center of new cluster is calculated by centroid method

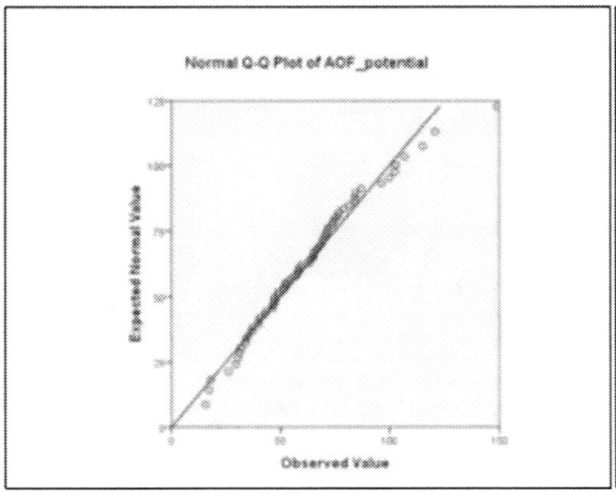
Normal Q-Q Plot of AOF_potential
Expected Normal Value
Observed Value

a.

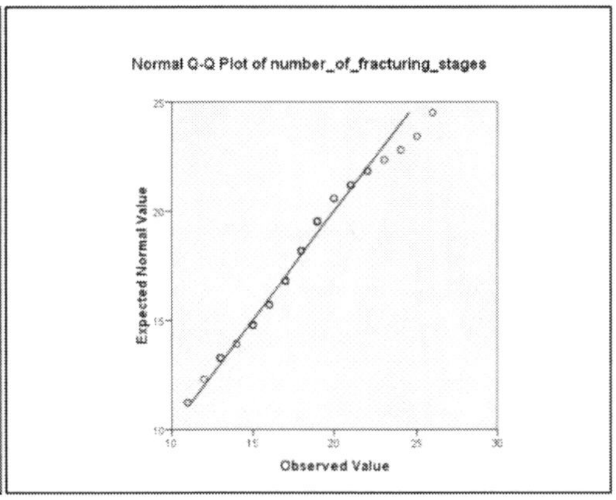
Normal Q-Q Plot of number_of_fracturing_stages
Expected Normal Value
Observed Value

b.

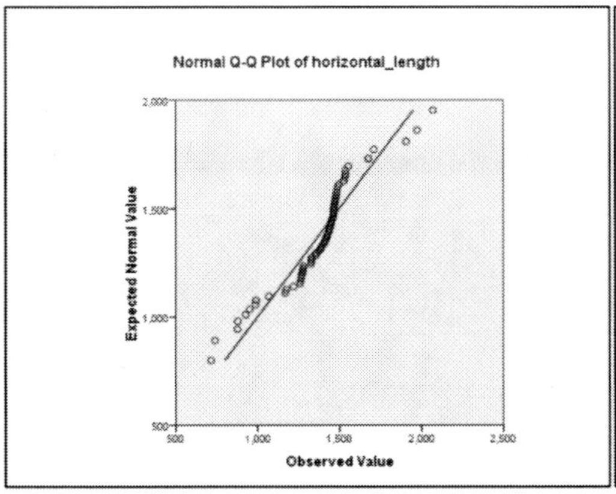
Normal Q-Q Plot of horizontal_length
Expected Normal Value
Observed Value

c.

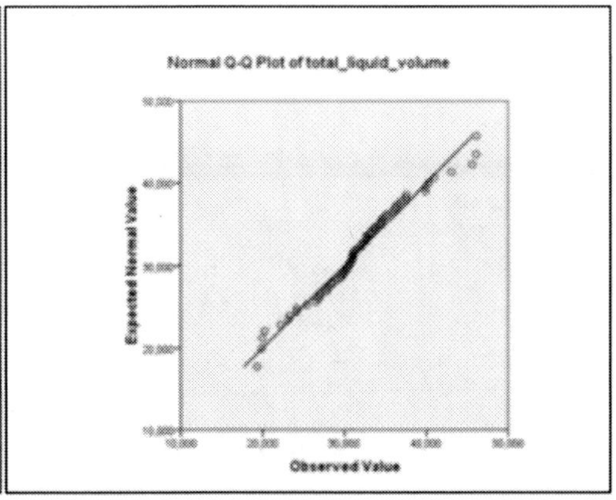
Normal Q-Q Plot of total_liquid_volume
Expected Normal Value
Observed Value

d.

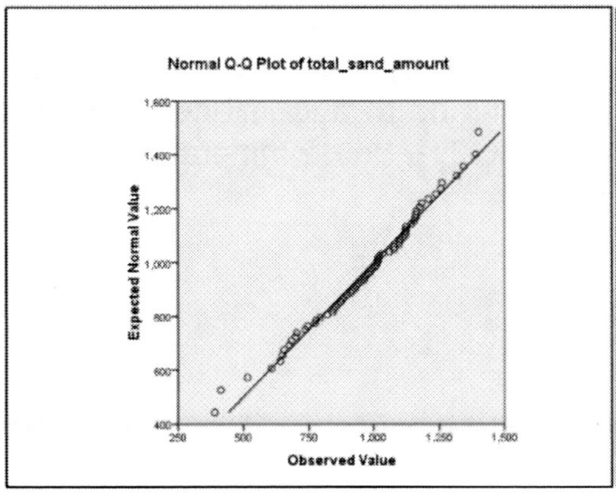
Normal Q-Q Plot of total_sand_amount
Expected Normal Value
Observed Value

e.

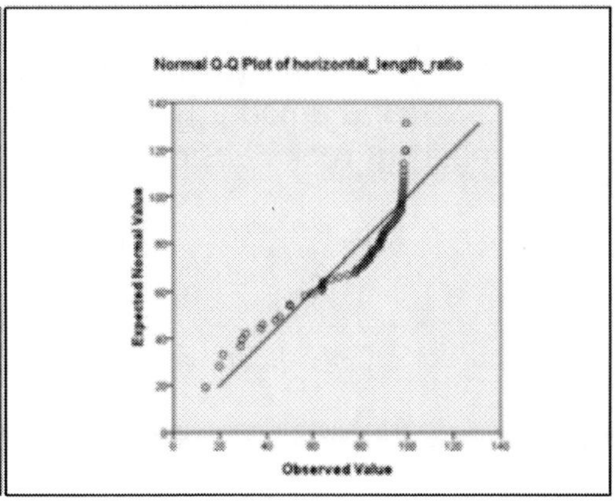
Normal Q-Q Plot of horizontal_length_ratio
Expected Normal Value
Observed Value

f.

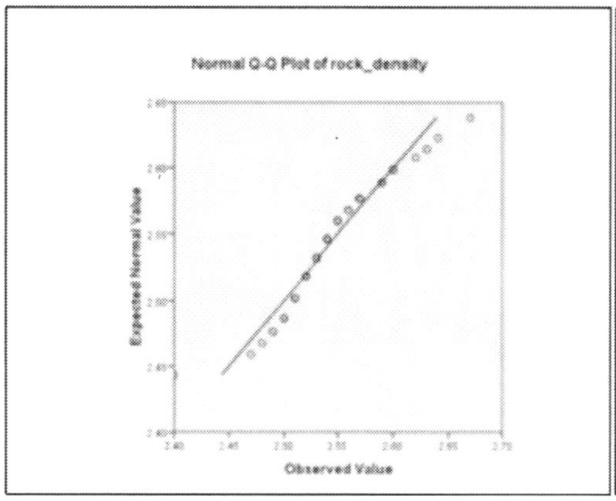

g.

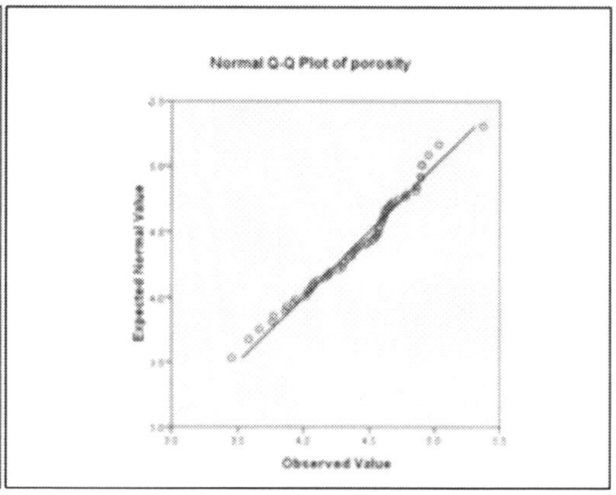

h.

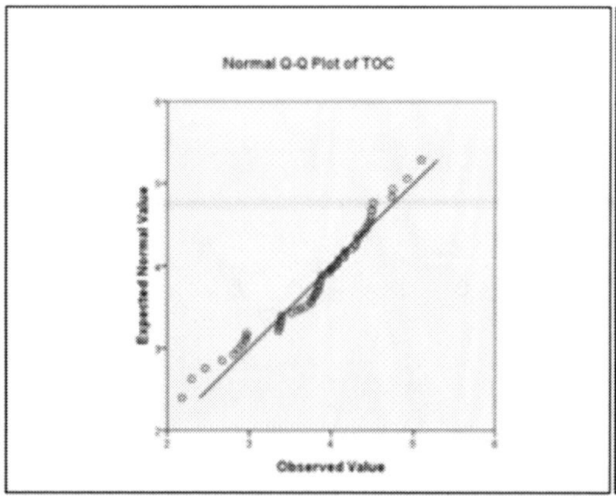

i.

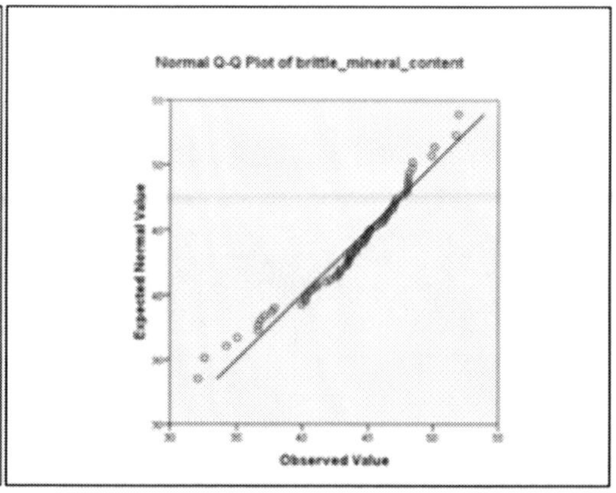

j.

Fig.4 The Q–Q (quantile-quantile) plots for the 10 variables of the 86 wells: a. AOF potential; b. number of fracturing stages; c. horizontal length of well; d. total liquid volume; e. total sand amount; f. horizontal length ratio through stage 1 and stage 3 of Long1 member; g. rock density; h. porosity; i. TOC; j. brittle mineral content of rock

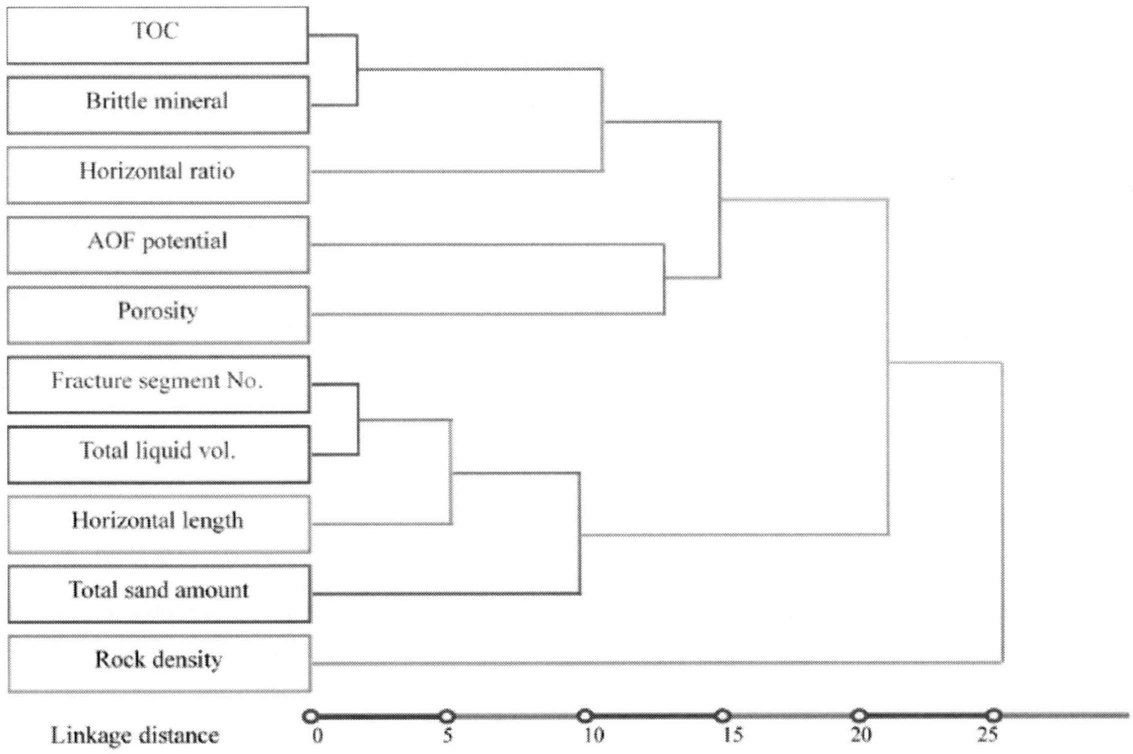

Fig.5 Hierarchical clustering analysis result of the sample data of 86 wells with linkage distance

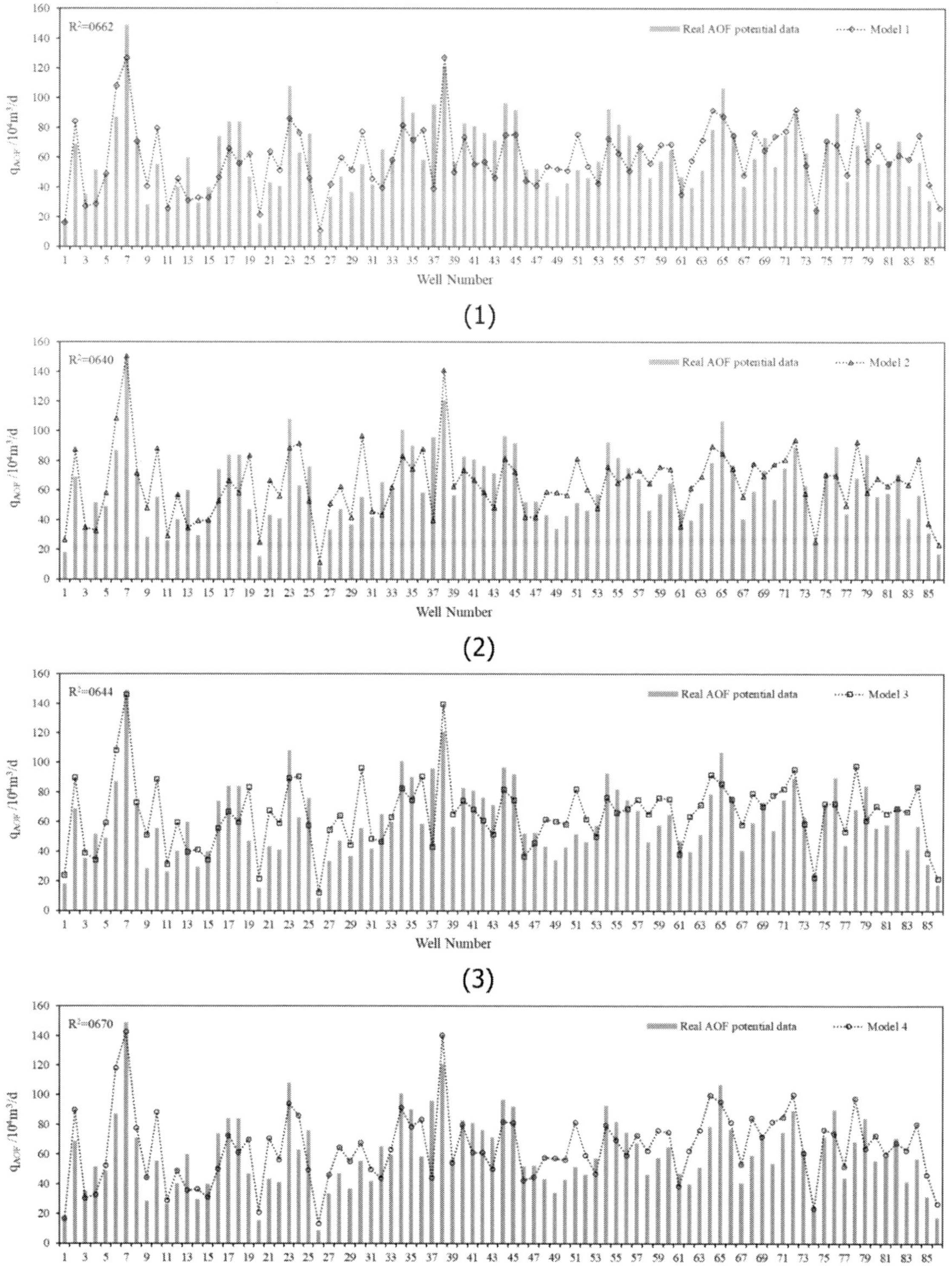

Fig.6 Comparison of actual and predicted AOF potential by the 4 models for the 86 wells with consideration of different key factors in Fuling Jiaoshiba shale gas field

Table 1 Correlation coefficients matrix of all variables in Jiaoshiba area

Factors	AOF potential	No. of fracturing stages	Horizontal length	Total liquid Volume	Total sand amount	Horizontal length ratio through stage 1 and stage 3	Rock density	Porosity	TOC	Brittle mineral content of rock
AOF potential	1	0.354	0.348	0.405	0.194	0.285	-0.373	0.246	0.433	0.438
No. of fracture stages	0.354	1	0.897	0.946	0.765	0.273	-0.203	-0.013	0.241	0.235
Horizontal length	0.348	0.897	1	0.863	0.633	0.331	-0.241	0.024	0.26	0.252
Total liquid Volume	0.405	0.946	0.863	1	0.736	0.302	-0.248	-0.01	0.274	0.282
Total sand amount	0.194	0.765	0.633	0.736	1	0.22	-0.002	-0.072	0.072	0.07
Ratio through stage 1 and stage 3	0.285	0.273	0.331	0.302	0.22	1	-0.492	0.287	0.568	0.578
Rock density	-0.373	-0.203	-0.241	-0.248	-0.002	-0.492	1	-0.289	-0.908	-0.917
Porosity	0.246	-0.013	0.024	-0.01	-0.072	0.287	-0.289	1	0.419	0.336
TOC	0.433	0.241	0.26	0.274	0.072	0.568	-0.908	0.419	1	0.982
Brittle mineral content of rock	0.438	0.235	0.252	0.282	0.07	0.578	-0.917	0.336	0.982	1

Table 2 Hierarchical clustering analysis result of factors in Jiaoshiba area

Parameter	Hierarchical clustering analysis result
AOF potential	1
Number of fracturing stages	2
Horizontal length	2
Total liquid volume	2
Total sand amount	2
Horizontal length ratio through stage 1 and stage 3	1
Rock density	3
Porosity	1
TOC	1
Brittle mineral content of rock	1

Table 3 Factors chosen in regression model according to hierarchical clustering analysis results

Regression model	Factors considered
1	a. Total liquid volume (X_1) b. Porosity (X_2) c. TOC (X_3)
2	a. Fracture segment No. (X_4) b. Porosity (X_2) c. TOC (X_3)
3	a. Fracture segment No. (X_4) b. Horizontal length ratio through stage 1 and stage 3 (X_5) c. TOC (X_3) d. Porosity (X_2)
4	a. Total liquid volume (X_1) b. Horizontal length ratio through stage 1 and stage 3 (X_5) c. Porosity (X_2) d. Brittle mineral content of rock (X_6)

601617 Solutions of C6F13- and C4F9-Based Polyfluoroacrylates in CO_2; Solubility, Viscosity, and Adsorption Onto Cement

Abstract:

High molecular weight poly(fluorinated acrylates) (PFAs) have drawn interest in the chemical and petroleum engineering fields due to their remarkable solubility in liquid and supercritical carbon dioxide (sCO_2) without the need for any co-solvent. This study does not include any examples of PFA based on monomers with the -C8F17 moiety because the ultimate hydrolysis degradation products of such polymers leads to the formation of PFOA, a compound deemed unsafe by the EPA. Instead, we have synthesized PFA based on fluoroacrylayte monomers containing the -C6F13 group (the functionality used in current commercial fluoroacrylate copolymers). These -C6F13-based polymers would degrade to form perfluorohexanoic acid (PFHxA) that, while significantly safer than PFOA, is also receiving more scrutiny from regulatory agencies in recent days. Therefore, we have also synthesized PFA based on monomers containing the -C4F9 functionality. The -C4F9-PFA ultimate degradation products include perfluorobutanoic acid (PFBA), which is significantly more environmentally benign than the -C6F13-based PFA. Therefore, one of the main objectives of this study was to determine if -C4F9-PFA retained the remarkable CO_2-solubility and CO_2-thickening capability as PFA based on the -C6F13 PFA and previously reported -C8F17-based PFA.

Many samples of PFA based on the -C6F13 and -C4F9-based fluoroacrylate monomers and varying ratios of initiator:monomer were bulk polymerized and then characterized for molecular weight and polydispersity. The -C4F9-PFA was shown to be nearly identical in CO_2-solubility as the -C6F16-PFA. Between 1-4 wt% PFA, the solubility of PFA in CO_2 was nearly independent of PFA molecular weight. Further, the cloud points loci for mixtures of CO_2-C4F9-PFA were nearly identical to the cloud point curves for the CO_2-C6F13-PFA at each temperature, with higher pressure required at higher temperature. This demonstrated that the -C4F9-PFA was just as CO2-soluble as the -C6F13-PFA. Further, both polymers were capable of thickening CO_2 in dilute concentration to a comparable degree (e.g. a 6-fold increase in viscosity with 4 wt% C4F9 PFA, 1.24×10^6 Da PFA in CO_2 at low shear rate).

PFA has previously been considered for use as a "CO_2 thickener" in enhanced oil recovery (EOR) operations. Unlike common chemical engineering applications where low viscosity is beneficial to reduce mass transfer resistance, low CO_2 viscosity is often problematic for EOR because it leads to viscous fingering (mobility control problems) and preferential loss of disproportionate amounts of CO_2 to high permeability "thief zones" that contain little recoverable oil (conformance control problems). However, in 2019 it was shown that PFA dissolved in CO_2 adsorbs so strongly onto sandstone or carbonate rock that the use of PFA in large-scale EOR projects to address mobility control would not be feasible, although near-wellbore conformance control applications (i.e. injecting the CO_2-PFA into isolated thief zones) may be viable.

In the current study, however, we are attempting to exploit the adsorption of PFA on solid surfaces for improved wellbore integrity. We are proposing to inject PFA-CO_2 solutions into small cracks within cement or between steel and cement (i.e. microannular cracks), thereby allowing a portion of the PFA to absorb and greatly reduce the crack permeability or completely seal the crack. Because PFA is a high molecular weight, amorphous, transparent, sticky, stretchy, oil-repellant, water-repellant polymer that is insoluble in water, oil or natural gas, it is a promising sealant for wellbore applications. Even though PFA is inherently expensive, this is a small volume application in which the cost of PFA would not exclude its consideration. Relative to other sealants (cement, resins, aqueous emulsions), PFA-CO_2 solutions are orders of magnitude less viscous, which would allow them deeper access into smaller cracks than other options. The PFA-CO_2 solutions would, however, be an inappropriate selection for the remediation of extremely large cracks or voids in cement.

In order to lay the groundwork for this proposed application, we have also determined the equilibrium adsorption of PFA onto Portland cement particles. This was accomplished by a direct measurement of the mass gain of a known quantity of sieved (22-30 mesh) Portland cement particles that was immersed in a PFA-CO2 fluid of specified original composition (e.g. 4 wt% PFA). The equilibrium concentration of PFA in the liquid CO_2 was determined via mass balance (original PFA placed in the cell – the amount of PFA found to adsorb onto the cement). This is, to the best of our knowledge, the first report of polymeric adsorption levels of solid surfaces immersed in CO_2-polymer solutions

611710 Competitive Sorption of CO2 in Shale Nanopores for Sequestration and Enhanced Gas Recovery Using Molecular Density Functional Theory

Abstract:

Enhanced gas recovery (EGR) by carbon dioxide (CO_2) injection to gas-bearing shale formations has been shown to be feasible and beneficial experimentally from pore scale to field scale. In addition to the improved recovery factor, CO_2 is trapped in the geological formation, which provides an opportunity to store CO_2. When supercritical CO_2 is injected, the preferential adsorption and dissolution of CO_2 in nanopores and organic matter enables the displacement of shale gas. This mechanism has been investigated by molecular simulation in graphite pores, but rarely for permeable pores by simulation or classical density functional theory (DFT). The compositional distribution of components from the organic matter source (kerogen) to nanoscale pores in the organic matter and to large fractures makes the estimation of fluid storage and the multicomponent compositional distributions in shale a challenging problem. We have introduced a density functional theory that accounts for molecular size and shape and kerogen maturity and swelling to predict the partitioning of components between the kerogen matrix and nano-pores in the kerogen [1].

In this work, a molecular thermodynamic model is presented that predicts both the adsorption of hydrocarbons in nanoscale pores and the dissolution of fluids in organic matter in equilibrium with the bulk kerogen phase. The swelling ratios of five kerogens of varying maturities and types in different solvents are well quantified using the new model. This kerogen matrix model is incorporated in interfacial statistical associating fluid theory (iSAFT) DFT to form a nanoporous kerogen composite model. The equilibrium partitioning of mixtures of fluids from the bulk phase to the nanoporous kerogen phase show enrichment of aromatic and cyclic molecules in kerogen, the preferential adsorption/absorption of high molecular weight molecules and carbon dioxide, and the significant amount of hydrocarbon storage both in pore space and kerogen matrix [1, 2].

Further, we apply molecular DFT to study the competitive sorption of CO_2 with shale gas in kerogen at various conditions. The DFT model is verified by grand canonical Monte Carlo (GCMC) simulation in graphite slit pores for pure and binary component systems at different temperatures, pressures, pore sizes and bulk compositions for methane/ethane/ CO_2. [2]. The model is utilized to predict the CO_2 adsorption selectivity in multicomponent systems which include two different shale gases. We show the selectivity of CO_2 decreases as temperature / pressure/ nanopore size/ shale gas average molecular weight increase. Subsequently, we model the CO_2/CH_4 adsorption in pores of more realistic conditions, in which we consider different levels of moisture in the pore and the gas dissolution in organic matrix. Interestingly, the presence of water inside a nanopore greatly reduces the CO2 adsorption but improves the CO_2 selectivity with respect to methane. The presence of organic matter kerogen also increases the CO_2 selectivity. In summary, a novel permeable pore DFT model enables prediction of partitioning and distribution of CO2 and hydrocarbons in shale related to enhanced gas recovery and CO_2 storage. [1, 2].

1. Jinlu Liu and Walter G. Chapman, "Thermodynamic Modeling of the Equilibrium Partitioning of Hydrocarbons in Nanoporous Kerogen Particles," Energy & Fuels, 33, 891-904 (2019).
2. Jinlu Liu, Shun Xi, and Walter G. Chapman, "Competitive Sorption of CO2 with Gas Mixtur in Nanoporous Shale for Enhanced Gas Recovery from Density Functional Theory," Langmuir, 35, 8144-8158 (2019).

609261 Spatial and Temporal Evolution of Natural Gas and Natural Gas Liquid Feedstocks in Shale Basins

Abstract:

Technological advancements in horizontal drilling and hydraulic fracturing have led to increasing production of light hydrocarbons from shale resources, including natural gas and natural gas liquids (NGLs). However, some basins that have insufficient gas gathering and pipeline networks (e.g. the Permian Basin in Texas) may flare significant quantities of gas. For example, in the Permian Basin, about 4.6% of gross gas production was estimated to be flared between May 2018 and March 2019 (Zhang et al., 2020). To mitigate flaring issues, hydrocarbon resources can be locally transformed to chemical or energy products. The National Science Foundation's Center for Innovative and Strategic Transformation of Alkane Resources (CISTAR) is developing modular processes to transform natural gas and NGLs to gasoline-like transportation fuels.

Determining the scale of operation of such modular chemical manufacturing requires a comprehensive understanding of the temporal and spatial availability of the hydrocarbon feedstocks. Preliminary work was performed in the Eagle Ford oil and gas producing region in south central Texas. With over 20,000 wells, the Eagle Ford produced 5.4 billion cubic feet natural gas per day and 1.2 million barrels of condensate and oil per day in 2019 (RRC, 2020). From the northwest to the southeast portion in the Eagle Ford Shale, well production varies from oil, to gas with condensate (wet gas), and to dry gas, as shown in Figure 1 (Gherabati et al, 2016). From oil to wet gas to dry gas regions, propane to ethane ratio varies from 1:1 to 1:3 (mass basis). In addition to varying gas compositions, the magnitude of available feedstocks also vary across the region. Ethane available at individual gas plants was estimated by assuming that produced gas from wells is sent to their nearest gas plant. Among 44 gas processing plants in the Eagle Ford, 60% of ethane production are represented by 43 plants with 200,000 – 750,000 kg of ethane per day, with one plant processing at ~3,000,000 kg/day, accounting for 40% of ethane production, shown in Figure 2.

Similar work has been performed in the Permian Basin in Texas. The Permian Basin is one of the largest oil-producing regions in the world and accounts for over 30% of oil production in the United States. Recent study (Zhang et al, 2020) shows that methane emissions in the Permian Basin are the highest among all the U.S. oil/gas producing regions. This presentation will provide detailed characterizations of spatial and temporal variability in light alkane production, and losses due to flaring for a case study region in the Permian Basin.

Collectively, the analyses in the Eagle Ford and the Permian Basin will provide two case studies of spatial and temporal characteristics of feedstock availability for modular light alkane manufacturing.

References

Gherabati, S. A., Browning, J., Male, F., Ikonnikova, S. A., & McDaid, G. (2016). The impact of pressure and fluid property variation on well performance of liquid-rich Eagle Ford shale. *Journal of Natural Gas Science and Engineering*, *33*, 1056-1068.

Railroad Commission of Texas (RRC) – Eagle Ford Shale Information. (2020). Available at: https://www.rrc.state.tx.us/oil-gas/major-oil-and-gas-formations/eagle-ford-shale-information/

Zhang, Y., Gautam, R., Pandey, S., Omara, M., Maasakkers, J. D., Sadavarte, P., ... & Zhang, R.

(2020). Quantifying methane emissions from the largest oil-producing basin in the United States from space. *Science Advances*, *6*(17), eaaz5120.

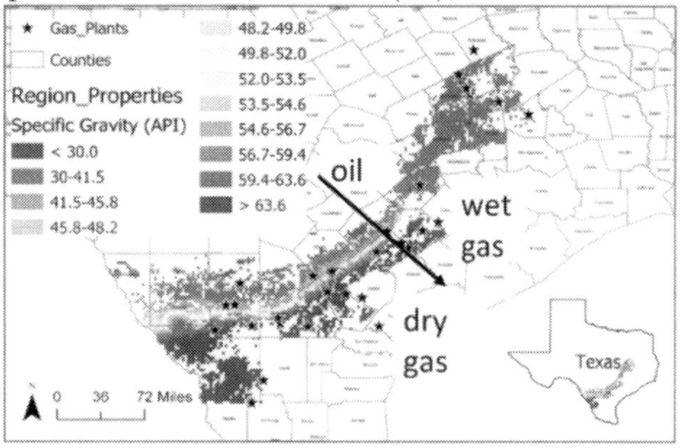

Figure 1. Spatial variabilities in well production in the Eagle Ford Shale.

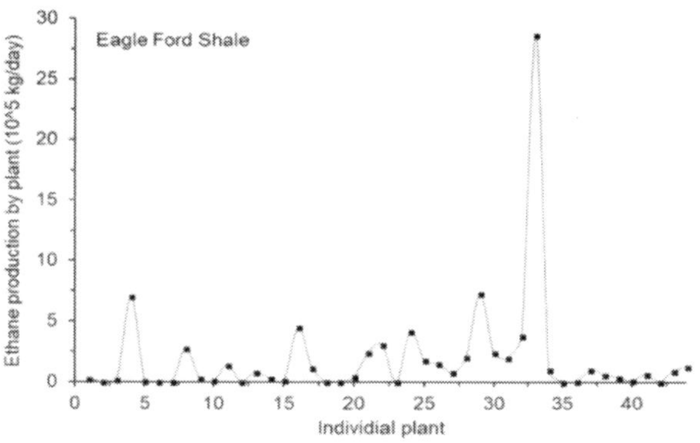

Figure 2. Ethane available at 44 gas processing plants in the Eagle Ford Shale

592706 A New Geology-Engineering Integration Solution for Shale Gas Exploration: A Case Study from Changning Block in China

Abstract:

Rich and quality shale gas reservoirs have been found in the Lower Silurian Longmaxi strata in the southern Sichuan Basin, which was identified as great development potential. However, shale gas reservoir is characterized by ultra-low permeability, which usually has no natural productivity and is called a typical “artificial gas reservoir”. We must pay attention to reservoir quality, drilling quality and completion quality in order to obtain the maximum EUR and the highest oil recovery. To tackle this challenge, the integrated Geoengineering solution was deployed in Changning Shale Gas Block, which consists of geological modeling, geomechanics, fracturing simulation and gas reservoir simulation. Geoengineering integration technology is used on the Changning A platform for field applications. The results show that: (1) Carrying out integrated research on geological engineering can realize the organic combination of geology and engineering. The geological model provides technical support for engineering optimization. The engineering practice constantly corrects the geological model and improves the precision; (2) Through the integration of geological engineering, we can create "transparent" shale gas reservoir, which can improve the predictive accuracy of geological and engineering sweet spot mapping; (3) Hydraulic fractures can be fine characterization by using geological engineering integration technology, which is useful for quantitative description and analysis of key parameters such as geometrical shape, net pressure, proppant distribution and conductivity of hydraulic fractures; (4) the refined unstructured-grid (<10m) based 3D numerical simulation model was built for hydraulic fractured horizontal wells in shale gas reservoir, the history match for shale gas wells has been hugely improved: gas production rate was matched to more than 98%; the wellhead pressure was matched to more than 80%; the cumulative liquid production rate was matched to more than 90%. With the implementation of the Geoengineering study, the shale gas production in Changning area has been increasing rapidly in the past 2 years.

593227 New Technology of EOR of Ultra Deep and Complex Condensate Gas Reservoirs in Tarim Basin， China

Abstract:

A number of condensate gas reservoirs (both sandstone and carbonate) with depth of over 5000m have been developed in Tarim Basin, China. Those reservoirs are featured with thick reservoirs pay zones (over 100 meters) with ,strong reservoir heterogeneities in both rock properties and fluid properties. Gas channeling in the process of gas injection development and retrograde condensation in the process of depletion development have seriously affected the recovery of condensate. In order to enhance the condensate recovery in those condensate gas reservoirs, a series of new technologies have been developed during the last decade.

Laboratory experiment analysis and development practice discovered that three major mechanisms controlling the gas migration process in the reservoirs: displacement, gravity and diffusion for cyclic gas injection. It is proposed that gravity differentiation is the main factor affecting the development effect of cyclic gas injection, and the fluid distribution law of underground condensate gas reservoir is re-recognized. For the sandstone condensate gas reservoir developed with large well spacing (over 1000 meters), the displacement and diffusion will be replaced by gravity differentiation gradually in the process of cyclic gas injection, and the fluid will be redistributed vertically according to its density, which makes different phase behaviors. Similarly, the reservoir space of depleted carbonate condensate gas reservoir is mainly dissolution cave and fracture. Under the action of gravity, the reverse condensate oil usually accumulates at the bottom of fracture cave system. On this basis of this understanding, the technology of gravity assisted three-dimensional displacement gas injection for sandstone condensate gas reservoirs and horizontal well placement technology for carbonate condensate gas reservoirs were developed to optimize the well placement. The technologies have been successfully applied to the developed condensate gas field in Tarim Basin, and the recovery rate of condensate oil has been increased by over 19%.

Production capacity evaluation method of condensate gas reservoir considering the influence of stress sensitivity and condensate oil

DU Chunxiao, LIU Xiaohong, ZHANG Fengyi, XU Jing, GENG Zhigang

(Tianjin Branch of CNOOC Ltd., Tianjin 300459)

Abstract: Accurate evaluation of production capacity is an important guarantee for efficient development of condensate gas reservoir. For the deep condensate gas reservoir with low permeability and low pressure difference between the initial formation pressure and the dew point, the influence of reservoir stress sensitivity and condensate oil on the production capacity of the gas reservoir cannot be ignored. Nevertheless, at present, the widely used method to calculate the gas reservoir production capacity is based on single-phase gas, without considering the influence of condensate oil and stress sensitivity on reservoir permeability. On the basis of researching a large number of literatures, combined with laboratory experiments and numerical simulation, the production capacity evaluation model of condensate gas reservoir under the comprehensive influence of stress sensitivity and condensate oil is derived by using the seepage flow mechanics theory, and the research results are applied to the Bohai Z gas reservoir. The results show that the production capacity evaluation model can more accurately reflect the production status of this kind of gas reservoir in different development stages, and can be widely used in the productivity evaluation of similar condensate gas reservoir.

Key words: deep condensate gas reservoir; stress sensitivity; condensate oil; evaluation of production capacity; low permeability

Gas reservoir production capacity evaluation is an important research content of gas reservoir engineering research, which provides key guidance for the design of gas reservoir development plan. For condensate gas reservoir, the fluid flow law and phase change are more complex, and the reverse condensation phenomenon has a serious impact on gas well productivity. According to theoretical and experimental studies, many scholars have confirmed that low permeability reservoirs have strong stress sensitivity. At present, the widely used method to calculate the gas reservoir production capacity is based on single-phase gas, without considering the influence of condensate oil and stress sensitivity on reservoir permeability. Many scholars have carried out research on the influence of reservoir stress sensitivity or retrograde condensation on gas well productivity, but lack of comprehensive consideration of both. Therefore, a new productivity evaluation method of condensate gas reservoir considering stress sensitivity and retrograde condensation is derived by combining laboratory experiments and numerical simulation methods, and is applied to BZ gas reservoir in Bohai oilfield.

1. Influence of stress sensitivity on reservoir permeability

The relationship between rock permeability and effective confining pressure has been studied deeply by many scholars using a large number of experiments. The relationship between permeability and effective confining pressure can be divided into three types: power-law relationship, exponential relationship and polynomial relationship.

BZ gas reservoir in Bohai oilfield is a low permeability gas reservoir. The core stress sensitivity test results of BZ-1 well are shown in Table 1 and figure 1. It can be seen that the dimensionless permeability of core decreases with the increase of effective stress, and BZ gas reservoir has strong stress sensitivity.

Table 1 Data of core stress sensitivity test

Net pressure	Sample 1	Sample 2	Sample 3	Sample 4
MPa	K_n/K_i	K_n/K_i	K_n/K_i	K_n/K_i
3.5	0.748	0.792	0.774	0.781
5.0	0.637	0.670	0.642	0.643
7.0	0.476	0.521	0.503	0.524
9.0	0.391	0.394	0.402	0.414
11.0	0.313	0.329	0.308	0.369
15.0	0.208	0.238	0.239	0.248
20.0	0.152	0.170	0.180	0.170
25.0	0.117	0.121	0.141	0.131
30.0	0.095	0.102	0.122	0.112
35.0	0.081	0.087	0.097	0.083

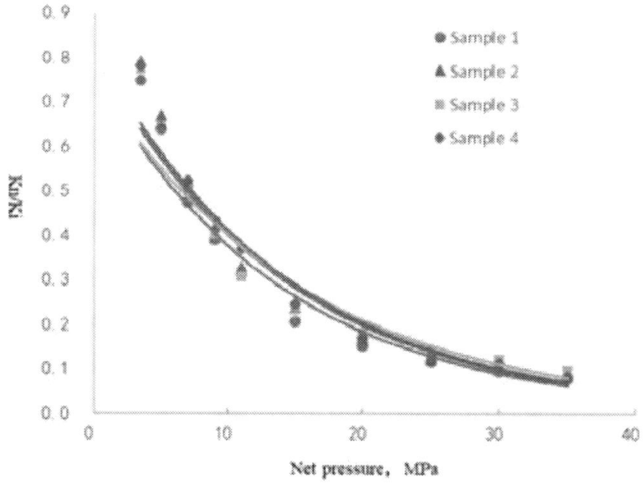

Fig 1 Stress sensitivity curves of the cores

By fitting the relationship between rock permeability and effective confining pressure, it is found that there is a good exponential relationship between them（Table 2）. It can be expressed as:

$$k_n = \mathrm{N}k_i e^{-\alpha(p_i - p)} \tag{1}$$

Table 2 Stress sensitivity index fitting of core permeability

Core number	Exponential fitting expression	R^2
1	$y = 0.7721e^{-0.072x}$	0.9553

2	$y = 0.8186e^{-0.071x}$	0.9595
3	$y = 0.7609e^{-0.064x}$	0.9489
4	$y = 0.8331e^{-0.07x}$	0.9727

According to the principle of seepage mechanics, the square distribution expression of stable seepage pressure of gas reservoir is as follows.

$$p^2 = p_e^{\,2} - \frac{p_e^{\,2} - p_{wf}^{\,2}}{\ln\left(\frac{r_e}{r_w}\right)} \ln\frac{r_e}{r} \tag{2}$$

Formula (1) is deformed.

$$k_n = Nk_i e^{-\alpha\left\{p_i - \left[p_e^{\,2} - \frac{p_e^{\,2} - p_{wf}^{\,2}}{\ln\left(\frac{r_e}{r_w}\right)} \ln\frac{r_e}{r}\right]^{\frac{1}{2}}\right\}} \tag{3}$$

2 Influence of condensate on gas phase relative permeability

The influence of reverse condensate oil is studied by numerical simulation method. By using single well numerical simulation, the variation law of condensate oil saturation at different distances from well during gas field development is obtained.

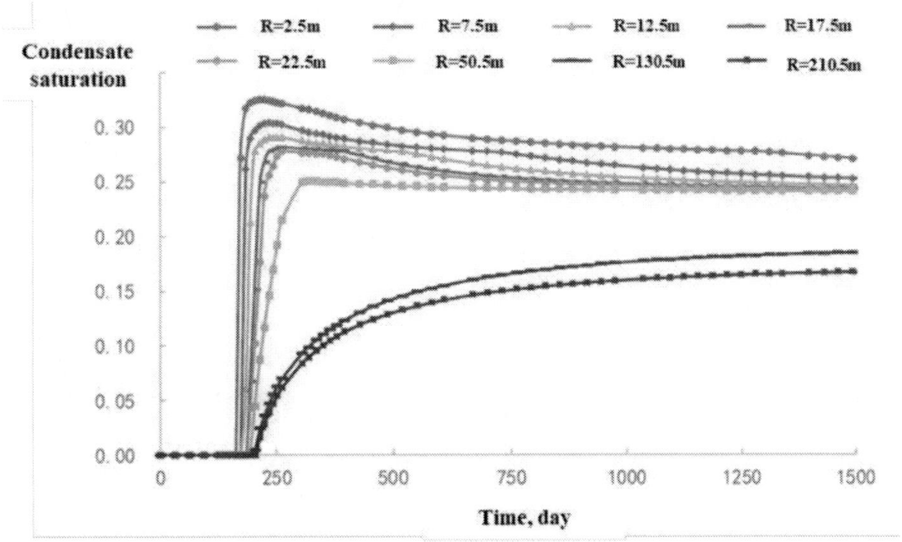

Fig.3 variation law of condensate oil saturation

It can be seen from Figure 3 that the condensate saturation decreases with the increase of radial distance, and tends to be stable after increasing with the extension of production time. The results show that the condensate saturation changes with time in the near well zone, and the distribution function of condensate saturation with radial distance $s_o(\mathrm{r})$ can be obtained at any time.

According to the relative permeability curve, the regression relationship between gas relative permeability and oil saturation can be obtained. The relative permeability curve of BZ gas reservoir in Bohai oilfield is shown in Fig. 4, the regression expression is:

$$k_{rg}(\mathrm{r}) = 4.7367s_o(r)^3 - 2.0497s_o(r)^2 - 2.1398s_o(r) + 1.0015 \tag{4}$$

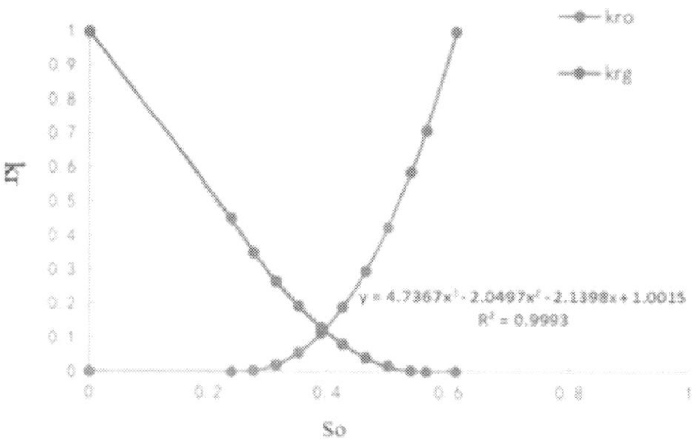

Fig.4 Relative permeability curve of Z gas reservoir

3 Improved gas well productivity evaluation model

The expression of effective permeability is:

$$k_g = k_{\mathrm{n}} \cdot k_{rg} \tag{5}$$

Replace formula (3) and formula (4) into formula (5):

$$k_g = Nk_i[4.7367s_o(r)^3 - 2.0497s_o(r)^2 - 2.1398s_o(r) + 1.0015]e^{-\alpha\{p_i - [p_e{}^2 - \frac{p_e{}^2 - p_{wf}{}^2}{\ln(\frac{r_e}{r_w})} ln\frac{r_e}{r}]^{\frac{1}{2}}\}} \tag{6}$$

The condensate saturation beyond a certain distance from the well is lower than the critical flow saturation, the distance is defined as the effective radial distance and expressed as r_o.

The whole stratum can be divided into two regions. The effective permeability can be expressed as:

$$k_g(\mathrm{r}, p_{wf}) = \begin{cases} Nk_a[4.7367s_o(r)^3 - 2.0497s_o(r)^2 - 2.1398s_o(r) + 1.0015]e^{-\alpha\{p_i - [p_e{}^2 - \frac{p_e{}^2 - p_{wf}{}^2}{\ln(\frac{r_e}{r_w})} ln\frac{r_e}{r}]^{\frac{1}{2}}\}}, & r_w < r < r_o \\ Nk_a e^{-\alpha\{p_i - [p_e{}^2 - \frac{p_e{}^2 - p_{wf}{}^2}{\ln(\frac{r_e}{r_w})} ln\frac{r_e}{r}]^{\frac{1}{2}}\}}, & r_o \le r \le r_e \end{cases} \tag{7}$$

① $r_w < r < r_o$, average permeability can be expressed as:

$$k_1(p_{wf}) = \frac{\int_{r_w}^{r_o} k_g(r, p_{wf}) \cdot 2\pi r dr}{\int_{r_w}^{r_o} 2\pi r dr} \tag{8}$$

② $r_o \le r \le r_e$, average permeability can be expressed as:

$$k_2(p_{wf}) = \frac{\int_{r_o}^{r_e} k_g(r, p_{wf}) \cdot 2\pi r dr}{\int_{r_o}^{r_e} 2\pi r dr} \tag{9}$$

According to the principle of seepage mechanics, the average gas permeability of the formation can be obtained:

$$k(p_{wf}) = \frac{k_1(p_{wf}) \cdot k_2(p_{wf}) \cdot \ln(\frac{r_e}{r_w})}{k_2(p_{wf}) \cdot \ln(\frac{r_o}{r_w}) + k_1(p_{wf}) \cdot \ln(\frac{r_e}{r_o})} \tag{10}$$

Based on the above research, the productivity evaluation formula of gas well considering stress sensitivity and condensate oil is obtained.

$$p_e^2 - p_{wf}^2 = \frac{1.273\times10^{-3}T\mu Z}{k(p_{wf})h}\ln\left(\frac{r_e}{r_w}\right) q_{sc} + \frac{2.162\times10^{-10}\gamma_g ZT}{r_w h^2 k(p_{wf})^{1.5}} q_{sc}^{2} \quad (11)$$

4 Application

BZ is a rare condensate gas reservoir with complex reservoir and fluid characteristics. The main performance is low pressure difference between the initial formation pressure and the dew point (0.21-7.13MPa), the maximum amount of condensate is 40%, the reservoir permeability is 1.00~6.80mD, strong stress sensitivity.

The evaluation equation is applied to BZ gas field. The IPR curves of three cases are calculated, under the comprehensive influence of condensate oil and stress sensitivity, only considering the influence of condensate oil or stress sensitivity. The results are compared with those of the traditional binomial formula.

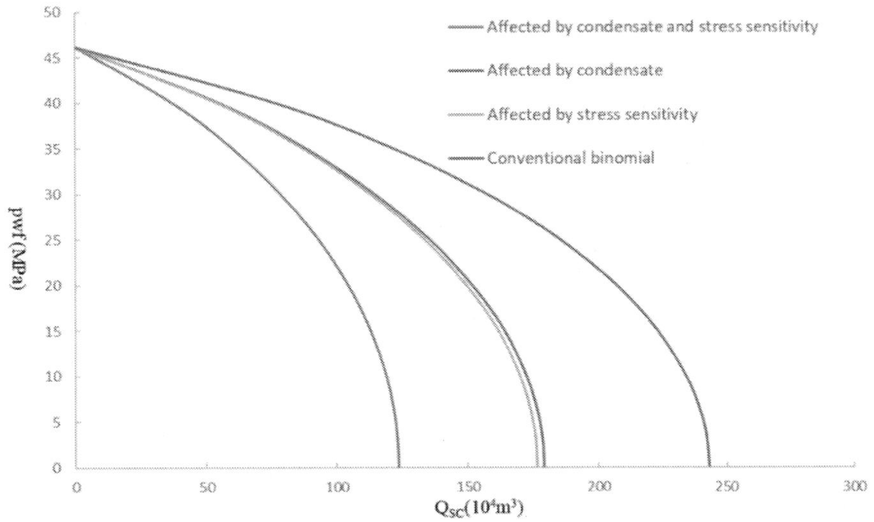

It can be seen that reservoir stress sensitivity and retrograde condensation have great influence on gas reservoir productivity. Under the combined effect of both, the productivity of gas reservoir is reduced by 50%. The IPR curves of gas wells under different formation pressures under the combined influence of stress sensitivity and retrograde condensation are further obtained.

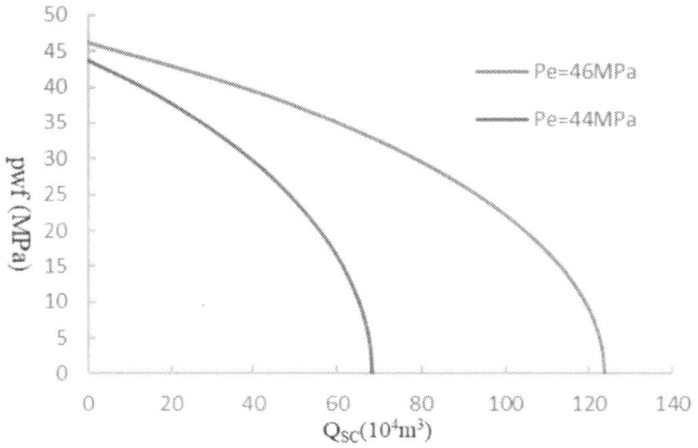

It can be seen that the formation pressure decreases, stress sensitivity and retrograde condensation increase, the productivity of gas wells also decreases obviously. The formation pressure drops by 2MPa, and the open flow rate of gas wells decreases by 57%. Therefore, in order to ensure the gas well productivity, it is necessary to adopt the development mode of gas injection and pressure maintaining, and the gas well production allocation should not be too high, in order to slow down the adverse impact of pressure drop.

6 Conclusion

(1) Stress sensitivity and condensate oil have great influence on the productivity of low permeability condensate gas reservoirs with low pressure difference between the initial formation pressure and the dew point. Most studies lack of comprehensive consideration of the two factors.

(2) Combined with laboratory experiment and numerical simulation, the formation permeability considering the effect of reservoir stress sensitivity and retrograde condensation is derived by using the principle of seepage mechanics, and then the evaluation method which can reflect the real productivity of gas reservoir is obtained.

(3) The research results show the necessity of gas injection to maintain reservoir pressure, and put forward suggestions for reasonable production of gas wells.

Symbol description

k_i-Original permeability, $10^{-3}\mu m^2$

k_n-Absolute permeability, $10^{-3}\mu m^2$

k_g-Effective permeability, $10^{-3}\mu m^2$

k_{rg}-Gas phase relative permeability, $10^{-3}\mu m^2$

p_i-Original formation pressure, MPa

p-Formation pressure, MPa

N-Stress sensitivity coefficient

α-Stress sensitivity index

p_e-Formation pressure at supply radius, MPa

p_{wf}-Bottom hole pressure, MPa

r_e-Supply radius, m

r_w-Wellbore radius, m

s_o-Condensate saturation near wellbore

r-Radial distance from well, m

T-Formation temperature, K

μ-Gas viscosity, mPa·s

Z-Gas deviation factor

h-Thickness of gas reservoir, m

β-Characteristic parameters of pore structure affecting turbulence and inertia resistance

γ_g-Relative density of natural gas

q_{sc}-Gas well production under standard conditions, m^3/d

References

[1] Fussel D.D. Single well Performance for gas condensate reservoir[J]. JPT，1973，25(12)：860-876.

[2] Jones J.R. and Raghavan R. Interpretation of Pressure buildup reservoirs in gas condensate well[J]. SPEFE，1989.

[3] FAN Xueping, XU Xiangrong. Experimental and mechanism research about permeability damage with the change of stress[J]. PETROLEUM EXPLORATION AND DEVELOPMENT, 2002

[4] XUE Zhan, LIU Pengcheng. A New Method of Correction of Gas Condensate Well Productivity Considering Capillary Number[J]. Journal of Chongqing University of science and Technology (NATURAL SCIENCE EDITION), 2014.

[5] LUO Ruilan, CHENG Linsong. A new method of determining relationship between permeability and effective overburden pressure for low-permeability reservoirs[J]. Journal of China University of Petroleum, 2007.

Numerical simulation study of a complex naturally-fractured gas reservoir with extra-high condensate in Bohai Oilfield

GAO Zhennan, SU Yanchun, GENG Zhigang, LIU Weilin, HUANG Baogang

(CNOOC China Limited Tianjin Branch, Tianjin 300459, China)

Abstract:

The BZ condensate gas field is the largest natural gas field in the history of the Bohai bay basin, with proven reserves of more than 100 billion cubic meters of natural gas and 100 million cubic meters of condensate oil. This gas field has unique features, such as deep burial, fracture development, strong stress sensitivity, small pressure difference between formation pressure and dew point pressure, high condensate content, etc. There is no similar condensate-gas field development cases in the world and proper reservoir numerical simulation study is crucial for efficiently developing such a reservoir.

This paper develops a numerical simulation flowchart for this kind of complex naturally fractured condensate gas reservoirs. First, based on logging, core and well testing data, the reservoir of BZ gas field is divided into three different types of deposits in the vertical direction: glutenite, semi-weathered zone and inner buried hill. The fluid flow among the three different types of deposits is calculated based on the equivalent fluid flow conduction between adjacent grids. Secondly, the correlation of permeability changes with stresses is established based on experimental data and implemented in the reservoir simulation based on the imbedded programming module in the simulator. Thirdly, through PVTi numerical simulation, EOS are tuned based on 7 pseudo-components. Finally, a dual-porosity compositional model is established, and the numerical simulation study is carried out to optimize the development plan. The developed flowchart will be a very useful tool for modeling naturally-fractured condensate gas reservoirs.

The simulation results indicated that: 1) considering the strong stress sensitivity, gas injection can effectively increase the permeability of the reservoir and control the condensate precipitation; 2) considering the structural characteristics of the reservoir, gas injection in the high position can form top-gas drive and expand the swept volume; 3) considering the influence of fracture, injection- production line perpendicular to fracture trend can effectively delay the breakthrough of injected gas and expand the injection capacity; 4) considering the characteristics of condensate, keeping formation pressure higher than dew point pressure can minimize the loss of condensate oil; 5) considering the mechanism of gas injection displacement, slug gas injection can form pressure pulse, expand the sweep volume and improve the recovery of condensate oil. Those findings can also be used to guide the development of other gas condensate reservoirs.

Key words: buried-hill fracture; condensate gas reservoir; stress sensitive characterization; reservoir numerical simulation; cyclic gas injection development

● 0 Introduction

BZ condensate gas field is the largest natural gas field in Bohai Sea, with proven natural gas reserves of over 100 billion cubic meters and proven condensate reserves of more than 100 million cubic meters. The gas field has the following geological characteristics: buried deep(-4400~-5500m), complex seepage (fracture development), stress sensitivity (permeability loss 89.2%), and the fluid characteristics are as follows: low pressure difference between formation pressure and dew point pressure (1.07MPa), ultra-high condensate (711g / m3), large amount of reverse condensate (32.3%). The efficient development of BZ condensate gas field is a world problem! It is found that there is no similar gas field development case at home and abroad, so it is impossible to accurately evaluate the development effect of this gas field. Therefore, it is urgent to carry out reservoir numerical simulation research to provide support for early development plan and subsequent efficient development[1-3].

● 1 Research method

According to the reservoir classification, the complex seepage system gas reservoir model is established. Based on the stress sensitive experiment, the variation law of seepage capacity is regressed. Based on the matching of fluid phase, the dual medium component model is established. And the development mode and gas injection development strategy are studied through reservoir numerical simulation.

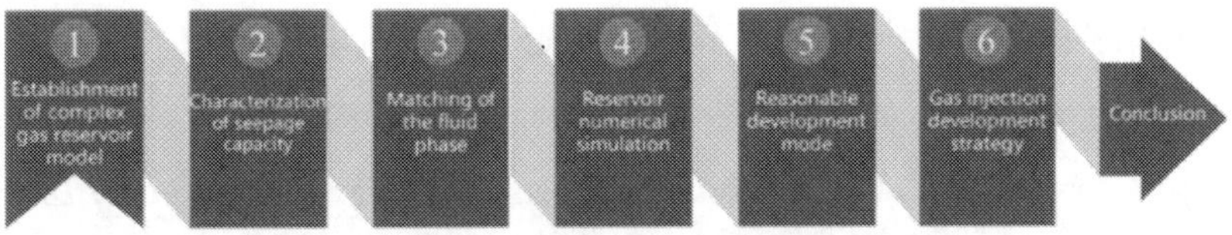

Fig 1. Technology Roadmap

● 2 Gas reservoir model

2.1 Reservoir zonation

Combined with the analysis of logging, core and testing data, BZ gas field has good reservoir connectivity, and it can be divided into sandy conglomerate zone, weathered zone and buried hill inner zone from top to bottom[4].

Fig 2. Vertical development characteristics of BZ condensate gas field reservoir

2.2 Establishment of complex gas reservoir model

Establish a structural model based on the understanding of reservoir zoning; and using lithofacies modeling and discrete fracture network modeling to generate matrix and fracture porosity, permeability and interporosity flow coefficient. By zeroing the matrix conductivity inside the buried hill, the coexistence of three types of complex seepage models in the BZ gas field can be realized.

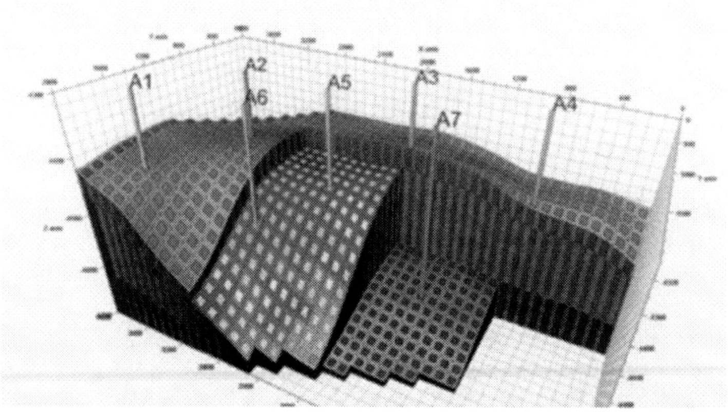

Fig 3. Gas reservoir model of complex seepage system

3 Stress sensitive characterization

3.1 Stress sensitivity analysis

According to the strong stress sensitivity of BZ gas field, the relationship between core permeability residual coefficient and net stress is fitted, and there is a good power function relationship between them; with the increase of net stress (reservoir pressure decreases), the rate of permeability loss gradually decreases[5-11]. In early stage, the net stress increased to 10MPa, Kr reduced by 60%; in mid stage, the net stress increased to 20MPa, Kr reduced by 20%; when reach the abandonment pressure(the net stress increased to 30MPa), the remaining permeability is about 10%.

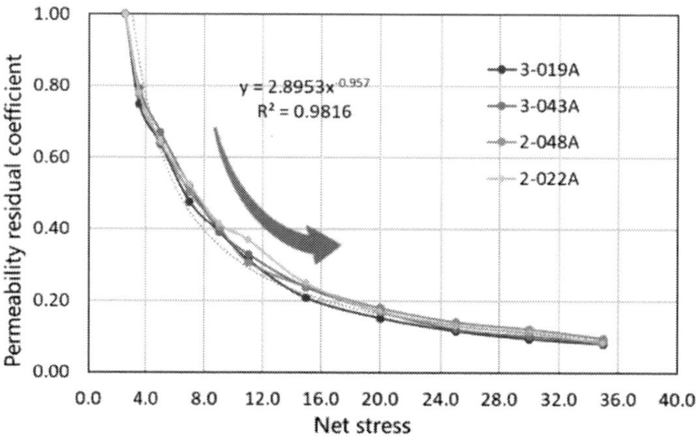

Fig 4. Permeability versus stress curve

$$K_r = 2.8953\ (P - P_0)^{-0.957}$$

K_r: Ratio of permeability to initial permeability, zero dimension;

P: Reservoir pressure at any time, MPa;

P_0: Initial formation pressure, MPa.

3.2 Stress sensitive characterization

According to the regression relationship, the characterization code of reservoir physical properties with time is compiled, and the sensitivity analysis is carried out by reservoir numerical simulation.

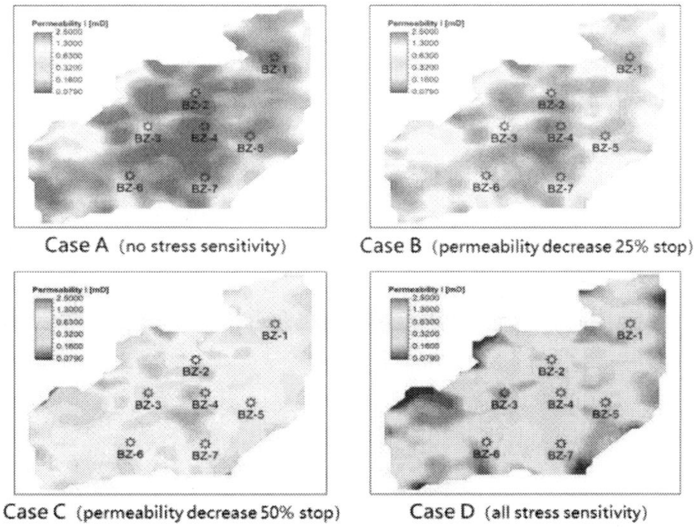

Fig 5. Comparison of permeability of different stress sensitive cases

• 4 Characterization of fluid characteristics

4.1 Fluid phase analysis

The BZ gas field is a condensate gas reservoir with high saturation and ultra-high condensate content through the study of fluid phase behavior characteristics, and the maximum condensate oil saturation is as high as 41%. When the pressure drops about 15MPa, the maximum saturation can be reached. Condensate oil precipitation seriously pollutes the reservoir, increases gas phase seepage resistance, and reduces gas well productivity and recovery[12-13].

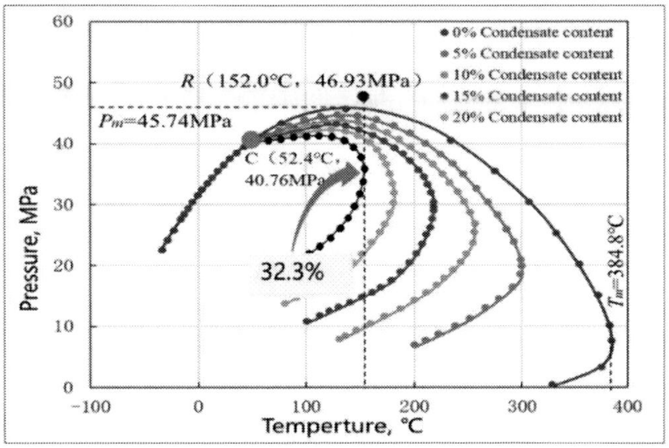

Fig 6. Phase diagram of fluid temperature and pressure in BZ gas field

4.2 Fluid phase matching

Considering the accuracy and speed of numerical simulation, 4 cases of component splitting combinations were set according to the principle of similarity combination. Combined with the calculation time and results of Numerical simulation of component model, the 7-component splitting plan is determined as the best plan.

11 components: C1、N2、C2、CO2、C3、C4、C5、 C6、C7-C10、C15+C22、C23+

7 components: C1+N2、C2+CO2、C3-C4、C5-C6、C7-C10、C15+C22、C23+

5 components: N2、CO2、C1、C2- C6、C7+

3 components: C1+N2、C2-C6+CO2、C7+

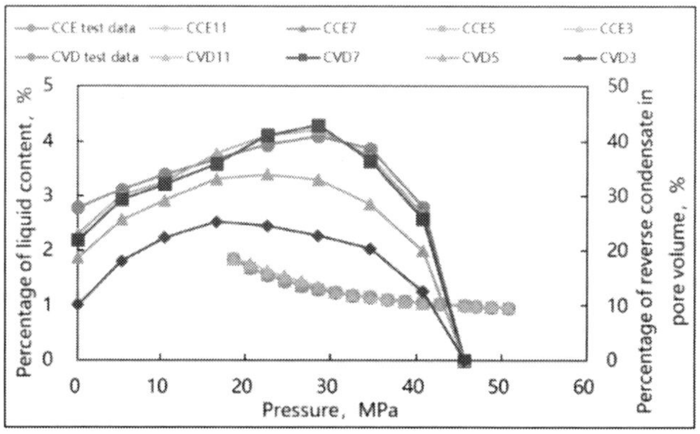

Fig 7. Comparison of experimental fitting accuracy

● 5 Study on gas reservoir numerical simulation

5.1 Optimization of development mode

As oil and gas fluids are produced, the formation pressure decreases, the physical properties become worse, the seepage ability decreases, and the recovery of gas field decreases. In depletion exploitation mode, the formation pressure drops rapidly, the condensate oil quickly reaches the maximum condensate velocity, and lost in the formation, so the condensate oil recovery rate is low. In gas injection development mode, the formation pressure is maintained through gas injection, and the retrograde condensate rate is controlled, and the condensate oil recovery is high. Considering reservoir stress sensitivity and retrograde condensate characteristics, from the perspective of improving oil and gas recovery and reservoir protection, it is recommended that the BZ condensate gas field development adopts first cycle gas injection and then depletion development[14-17].

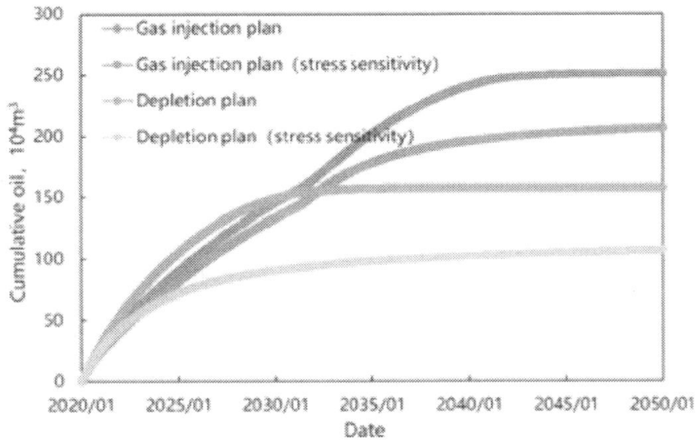

Fig 8. Comparison of different development mode

5.2 Cycle gas injection development strategy

Condensate gas reservoirs are usually developed with gas injection. The main purpose is to supplement formation energy, reduce condensate oil precipitation, and improve condensate oil recovery. When evaluating the effect of development, gas injection sweep volume and formation pressure maintenance level are two very important factors.

5.2.1 Optimization of injection and production

During gas injection development, the gas injection well position is directly related to the dry gas sweep volume. The structural and the fracture trend are two important factors.

A. Structural factor

The sandy conglomerate zone in structural high area of the BZ gas field has good physical properties. Priority is given to production in the early stage, and gas injection at the low area. In later stage, the condensate oil is enriched in the structural low area by gravity, and the top gas injection forms secondary gas and top drive. Vertical displacement, effectively expanding the gas sweep volume.

Tab 1. Development effect of structural factor

Plan	Cumulative gas 10^8m^3	Gas recovery %	Cumulative oil 10^4m^3	Oil recovery %
Inject in low area 10 years	36	53	181.3	26.6
Inject in low area 5 years,And in high area 5 years	36.6	53.3	192.4	28.3
Inject in high area 10 years	36.1	53.1	189.2	27.8

B. Fracture trend factor

When the injection-production line is parallel to the fracture trend, it is easy to form gas channeling, which reduce oil and gas recovery. When the injection-production line is vertical to the fracture trend, it can maximize the sweep volume, and effectively maintain formation energy and the seepage ability, and inhibit condensate oil to separate out.

Tab 2. Development effect of fracture trend factor

Plan	Cumulative gas $10^8 m^3$	Gas recovery %	Cumulative oil $10^4 m^3$	Oil recovery %
The injection-production line isparallel to the fracture	35.1	51.1	168.6	25.1
The injection-production line isperpendicular to the fracture	36.6	53.3	192.4	28.3
The injection-production line isoblique to the fracture	36	53	186.1	27.3

5.2.2 Reasonable pressure maintenance level

The most important thing in the development of condensate gas reservoirs is control the formation pressure to slow down the precipitation of condensate oil. The timing of gas injection and the rate of reinjection are two key factors.

A. Timing of gas injection

The timing of gas injection determines the degree of condensate oil loss, seepage capacity, and condensate precipitation rate during gas injection development. Since the difference between the formation pressure and the dew point pressure of the BZ gas field is very small, gas injection at the beginning of production can effectively delay the retrograde condensation time and reduce pollution, which is the optimal solution.

Tab 3. Development effect of timing of gas injection

Plan	Cumulative gas $10^8 m^3$	Gas recovery %	Cumulative oil $10^4 m^3$	Oil recovery %
Formation pressure 30MPa (4 years later)	26.4	48	127	23.4
Formation pressure 40MPa (2 years later)	26.3	47.9	132	24.4
Formation pressure 50MPa (0years later)	26.3	47.9	135.9	25.1

B. Re-injection rate

The higher the gas re-injection rate, the more stable the formation pressure remains. Before the formation pressure is higher than the dew point pressure, the 100% re-injection rate can always maintain the formation pressure higher than the dew point pressure, minimize the loss of condensate oil, and increase condensate oil recovery rate.

Tab 4. Development effect of Re-injection rate

Plan	Cumulative gas 10^8m^3	Gas recovery %	Cumulative oil 10^4m^3	Oil recovery %
100% re-injection rate	36.2	52.7	198.5	29.2
80% re-injection rate	36.4	53	196.1	28.8
50% re-injection rate	36.6	53.3	192.4	28.3

5.2.3 Optimization of gas injection mode

Considering the mechanism of gas injection and the demand for gas supply in the Beijing-Tianjin-Hebei region in winter, a comparative study of 3 plans: cyclic gas injection, slug gas injection, and Inject gas and soak. Circulating gas injection is prone to gas channeling, thereby reducing the condensate oil and gas recovery rate. The plan of inject gas and soak delays the gas channeling time on the basis of maintaining formation pressure, so the recovery rate is higher. In slug gas injection plan, the pressure pulse formed by the gas injection and gas production cycle can effectively expand the swept volume. And when injection stopped, the condensate in matrix will flow into the fracture due to the gravity. Therefore, the slug gas injection plan has the highest oil and gas recovery.

Tab 5. Development effect of Optimization of gas injection mode

Plan	Cumulative gas 10^8m^3	Gas recovery %	Cumulative oil 10^4m^3	Oil recovery %
Cyclic gas injection	35.3	50.2	188.4	26.8
Slug gas injection	37	53.9	198.4	29.1
Inject gas and soak	36.5	52.3	194.7	28.2

6 Conclusion

(1) The BZ gas field are divided into sandy conglomerate zone, weathered zone and Buried hill inner zone. The gas reservoir model of complex seepage system is established by the zeroing matrix conductivity inside the buried hill.

(2) The BZ gas field has strong stress sensitivity, and Nearly 90% loss of seepage capacity under abandonment pressure. Therefore, it is necessary to carry out numerical simulation research through time-varying reservoir physical properties.

(3) BZ gas field is a gas field with high saturation and high condensate content. Through PVTi matching and numerical simulation sensitivity analysis, the 7-component splitting plan is determined as the best plan.

(4) The numerical simulation study of the BZ gas field shows that: firstly inject gas in low area of the structure and then in the high area, inject gas at the start of production, the injection rate is 1.0, and seasonal slug gas injection is the optimal plan. This research provides strong support for

ODP program preparation and subsequent efficient development.

Reference:

[1] Fang Na, Jiang Guanghong, Li Yunpeng, et al. Development characteristics and potential tapping direction of buried-hill fractured oil reservoirs[J]. Special Oil & Gas Reservoirs, 2017,24(3):90-94.

[2] Xu Chunqiang, Zhang Zhen, Zhang Xintao, et al. Tectonic evolution and hydrocarbon accumulation model of western sag in western Bohai sea[J]. Journal of Northeast Petroleum University, 2018, 42（1）: 68-76.

[3] Zhang Feifei. Characterization and evaluation of pore-throat structures in ultra-low permeability and tight sandstones and their impacts on petrophysical properties: a case study of the lower Cretaceous in northern Songliao Basin[J]. Journal of Northeast Petroleum University, 2018,42(5):54-62.

[4] Tong Kaijun, Zhao Chunming, Lv Zuobin, et al. Petroleum exploration and development[J]. Petroleum Exploration and Development, 2012,39(1):56-63.

[5] Lin Guangrong, Liu Xuegang, Lu Yan, et al. Impact of stress change on percolation pattern of two-phase in low permeability reservoir[J]. Special Oil & Gas Reservoirs, 2008,15(1):43-45.

[6] Yang Bin, Hou Dali, You Gang, et al. Research and application of stress sensitive experiment on low permeable reservior[J]. Reservoir Evaluation and Development, 2012,5(2):41-45.

[7] Cao nai, Lei Gang. Stress sensitivity of tight reservoirs during pressure loading and unloading process[J]. Petroleum Exploration and Development, 2019,46(1):132-138.

[8] Jiang Ruizhong, Huang Lei, Cui Yongzheng, et al. Transit pressure analysis of a horizontal well in fractured composite reservoirs with stress-sensitivity effect[J]. Journal of Northeast Petroleum University,2018,42（3）: 84-91.

[9] Fang Yang. Research on relationship between deliverability of gas reservoir and the stress-sensitiveness of the reservoir[J]. Well Testing,2009,18（6）:18-21.

[10] Fu Lanqing. The effect of fractures on stress sensitivity and seepage characteristics in sandstone reservoirs[J]. Journal of Yangtze University（Natural Science Edition）,2016,13（20）: 14-18.

[11] Zhang Rui, Ning Zhengfu, Zhang Haishan, et al. New insights and discussions on stress sensitivity of fractured tight reservoir[J]. Natural Gas Geoscience, 2016,27(5):918-923.

[12] Wang Lezhi, Qi Zhilin, Jiang Yiwei. Study on retrograde condensate characteristics of deep condensate gas reservoirs with low permeability [J]. Natural Gas Technology, 2007, 1（1）: 52-54.

[13] Xiao Lixian, Du Jianfen, Guo Ping, et al. Pollution caused by retrograde condensate in condensate gas reservoir[J]. Fault-block Oil & Gas Field, 2009, 16（4）: 102-104.

[14] Cheng Liang, Zhang Xiaodong, Wang Juan, et al. Study on periodic gas injection mechanism of condensate gas reservoirs with high to high-super content of condensate oil in central Tarim by numerical simulation[J]. Journal of Xi'an shiyou University (Natural Science Edition),2016,31(2): 57-64.

[15] Wang Jing, Liu Huiqing, Liu Renjing, et al. Numerical simulation for low-permeability and extra-low permeability reservoirs with considering starting pressure and stress sensitivity effects[J]. Chinese Journal of Rock Mechanics and Engineering, 2013,32(2):3317-3327.

[16] Li Yong, Zhang Jing, Wu Xueyong, et al. A new reservoir simulation approach for fractured gas-condensate reservoirs[J]. Petroleum Exploration and Development, 2010,37(5):592-595.

[17] Jiang Hai，Yang Zhaozhong，Hu Yuehua，et al. Optimization of the cyclic gas injection parameters in condensate gas reservoir[J]. Journal of Xi'an shiyou University (Natural Science Edition),2009,24（2）:54-56.

611771 Characteristics of Phase Changing Material Nanoemulsions

Abstract:

Phase change material (PCM) is an energy storage alternative that consists of heat storage performance due to latent heat capacity. In buildings, PCMs can be utilized to save energy in heating and cooling systems implementing it into the walls of the structure. However, an alternative is circulating the PCM throughout the building by firstly turning them into emulsions. The motive is to create nanoemulsions of these PCMs with the lowest possible viscosity to ensure pumpability while increasing the amount of PCM by mass to enhance larger latent heat storage capacity which can reduce energy cost. The concern with nanoemulsion PCMs is choosing the correct emulsifiers as well as phase changing material based on applicability, stability, and economical costs. Knowing the finer details of these substances can be a stepping stone to choosing the correct materials.

This study aims to analyze the thermal performance and molecular structures of these PCM nanoemulsions through using various surfactants, thermal cycling, and the effect of aging for a size distribution ranging from 100 to 400 nm acquired by Dynamics Light Scattering (Malvern Zetasizer ZS90). Through rheological properties and molecular structures by NMR spectroscopy on these PCM nanoemulsions we may see how these nanoemulsions evolve and how this can create differences in the heat transfer coefficients obtained in forced convection through a heated circular tube. Octadecane is used as the PCM in the dispersed phase, with the utilization of different surfactants: stearic acid and the Tween & Span families, within the continuous phase, water. The preliminary data acquired of octadecane at a volume fraction of 30% with Span 60 and Tween 60 emulsifiers (3 wt%) showed shear thinning effects with the presence of minute yield stresses observed at all temperatures (10 -35°C) indicating a viscoelastic region; evident by the exponentially decreasing slope of the viscosity profile with an increase in shear rate (0.01-1000 S^{-1}). The temperature dependency is undoubtedly visible as the viscosity decreases with increase temperature from measurements performed on 50mm parallel plate geometry with a peltier heating plate for the higher volume fractions of oil (octadecane), as well as through a cup-and-bob geometry utilizing the ARES G2 Tensiometer (TA instruments). Solution-state ^{1}H and ^{13}C NMR measurements reveal molecular-level information on the local environments, compositions, and dynamics of the oil and surfactant species, which significantly depend on the thermodynamic state, age, and size distribution of the dispersed phase. Differential Scanning Calorimetry (TGA Q1500) measurements show the PCM nanoemulsion melting and freezing temperatures and how they change with thermal cycling and aging. A test section consisting of a 304 stainless steel tube heated uniformly and thermocouples installed at the inlet, outlet, and several on the tube wall to measure the temperature along the tube, the heat transfer coefficient can be acquired. This can indicate the possibility of using the PCM nanoemulsions as a secondary refrigerant for thermal energy storage and transport.

603005 Molecular-Level Understanding of Phase Behavior in Phase-Change Nano-Emulsions for Thermal Energy Storage

Abstract:

Phase Change Materials (PCMs) are latent heat storage materials that can store or release large quantity of energy while undergoing thermodynamic phase transitions. Organic PCMs are low cost, melt congruently, and exhibit good nucleating properties. To enable control over mixture transport properties, enhance thermal conductivity, and reduce cost, organic PCMs can be emulsified in water in the presence of surfactants to form PCM nano-emulsions. However, PCMs nano-emulsions can lose phase stability due to repeated melting and freezing processes during thermal cycling, which can further be deteriorated by shear flow during their applications in heat transfer systems. The distribution and dynamics of surfactants, and how they respond upon thermal cycling and shear, are poorly understood, but are expected to correlate with the observed phase, momentum, and heat transfer instabilities during their use.

To better understand the molecular-level origins of phase stability in PCMs nano-emulsions, multi-dimensional ^{1}H and ^{13}C solution-state nuclear magnetic resonance (NMR) methods have been applied to a molecularly simple octadecane-water-stearic acid system under thermal cycling. Octadecane freezing and melting were readily observed, yielding information on octadecane supercooling effects. ^{1}H pulsed-field-gradient (PFG) NMR methods were applied to measure the diffusion coefficients of the octadecane nano-emulsions, enabling estimates of their average sizes by the Stokes-Einstein equation. Chemical shift, relaxation, and diffusion of the component species were examined as indicators of phase stability as a function of temperature and number of thermal cycles. The results lay the groundwork towards molecular understanding of phase stability under thermal cycling in oil-water-surfactant phase-change nano-emulsions.

Effects of Operating Conditions for a Silica Gel/Water Vapour Adsorption Thermal Energy Storage System

C. Strong, Y. Carrier, F. H. Tezel

Abstract

Thermal energy storage technology can be used to further integrate renewable energy sources into the current global energy portfolio to reduce/eliminate the energy supply and demand mismatch. However, conventional thermal energy storage processes often require toxic materials or occupy large volumes to store useful amounts of energy. Therefore, thermochemical energy storage systems, which are potentially less hazardous and more compact than conventional thermal storage techniques, should be studied and developed further.

In this study, a water-adsorption based thermochemical energy system is analyzed. Thermal energy is stored when water vapour desorbs from the surface of the adsorbent by heating the material and/or using a purge gas. The energy is then released by exposing the adsorbent to humid air. The adsorbent used in this study was silica gel, due to its favourable water-adsorption properties, commercial availability, and cyclical stability. Thermocouples and relative humidity probes were used to measure the temperature and humidity at the inlet and outlet of the column. This, along with the measurement of flow rate allowed the calculation of the amount of energy released based on the temperature change from the inlet to the outlet of the column.

Various operating parameters like particle size, flow rate, desorption temperature, and relative humidity were manipulated in order to observe their effects on system energy storage performance variables like energy storage density, thermal power, and temperature lifts. The effects of these operating parameters are quantified and discussed herein, and the energy storage system was subsequently optimized.

Particle size in the tested range appeared to have minimal impact on performance. Increasing the desorption temperature had positive effects until approximately 120°C, beyond which significant increases in performance were not observed. Higher relative humidity levels during adsorption were shown to drastically improve system performance, due to the greater water-adsorption capacity of silica gel at high humidity levels. Finally, the optimal flow rate was found to be around 20-24 SLPM. An optimum was likely observed due to the competing factors of pressure, residence time, and mass transfer, which are all affected by the flow rate. Ultimately, it was found that the system performed the best when the silica gel was desorbed at 120°C, the flow rate during adsorption was 24 SLPM and the relative humidity during adsorption was 90%. Under these conditions, an energy storage density of 200.7 kWh/m^3, a maximum temperature lift of 28.5°C and a maximum thermal power of 245.2 W/L was observed.

Keywords

Thermochemical heat storage	Silica gel
Thermal energy storage	Operating conditions
Open sorption system	Optimization

601013 Experimental Investigation of a High-Temperature Thermochemical Storage Reactor

Abstract:

The integration of thermal energy storage to concentrated solar power plants enables round-the-clock dispatchability of industrial process heat and/or electricity. State-of-the-art molten salt storage systems are typically limited to applications at below about 600°C. Higher operating temperatures are desirable for efficient power generation and for driving energy-intensive thermochemical processes such as minerals and metals extraction, cement manufacturing, and fuels production. In particular, thermochemical heat storage (TCS) employs reversible endothermic/exothermic reactions to store heat with relatively high energy density and for long-term and seasonal storage. In the case of gas-solid reactions, the reaction equilibrium can be adjusted by controlling pressure and/or temperature, therefore controlling the uptake/release of heat.

We report on the design, fabrication, and experimental investigation of a TCS reactor applied for gas-solid reactions. It consists of an Inconel tubular shell filled with a packed bed of the solid reactive material. A concentric inner porous ceramic tube allows the gaseous reactant and product to flow radially through the packed bed with minimal pressure drop. A numerical heat and mass transfer model of the reactor has been developed for elucidating rate-controlling mechanisms and determining optimal design parameters. Model validation is accomplished with experimental data obtained with a lab-scale reactor prototype using the reversible calcination/carbonation of limestone as a model reaction: $CaCO_3 \leftrightarrow CaO + CO_2$. Powder and granules of $CaCO_3$ containing MgO (42 wt%) as an inert stabilizer exhibited stable cyclic performance for up to 100 consecutive carbonation-calcination cycles performed in a thermogravimeter analyzer (TG). During the TG measurements, a pure CO_2 environment at atmospheric pressure was maintained while swinging the temperature between 825°C (carbonation onset) to 925°C (calcination onset). The reaction extent was 96.48% for powder but only 31.99% for granules because of sintering at 1260°C during manufacturing. Similarly, the TCS reactor setup is operated in a temperature-swing mode at ambient pressure, which is ensured by storing the evolved gas (CO_2) in an inflatable bag during calcination and measuring the inflow/outflow of CO_2. The gas atmosphere is controlled by repeated evacuation and re-filling of the setup, and further verified by gas chromatography. The reactor tube contains 0.37L of the reactive material and multiple thermocouples for measurement of the radial temperature distribution, which in turn validates the transient conduction-convection-radiation heat transfer model within the packed bed. The engineering details of the TCS reactor as well as model and experimental results will be presented.

AUTHOR INDEX

Agarwal, Harsh ... 14
Ahmed, Luay ... 52
Al-Rubaye, Haider ... 67
Alembath, Anand ... 67
Alenzi, Adel ... 110
Allen, David ... 113
Alptekin, Gokhan ... 73
Aras, Chinmay ... 24
Baker, Sarah ... 17
Balbuena, Perla B. ... 44
Banerjee, Sanjoy ... 54, 56
Baratsas, Stefanos G. ... 23
Barteau, Mark A. ... 13
Barton, Paul I. ... 43
Bazant, Martin Z. ... 11
Beck, Victor A. ... 17
Berlinger, Sarah A. ... 50
Bhattacharyya, Debangsu ... 24
Bhosale, Rahul ... 72
Black, Brenna ... 61
Bonnema, Michael ... 73
Brown, David ... 6
Brown, Robert C. ... 62
Buchanan, Cailin ... 59
Cao, Yuan ... 13
Carrier, Ye ... 136
Cetinbas, Firat ... 47
Chaconas, Demetrios ... 26
Chang, Cheng ... 115
Chapman, Walter G. ... 112
Chen, Jee-Jay J. ... 13
Chen, Mengjie ... 46
Chen, Qining ... 113
Choi, Hoon ... 61
Choi, Hyun-Kyu ... 19
Chowdhury, Anamika ... 50
Colclasure, Andrew ... 6
Cottrill, Anton L. ... 58
Crandall, Dustin ... 110
Curtis, Strong ... 136
D'Ambrose, Michael ... 54
Dagan, Nick ... 50
Daniels, Benjamin ... 52
Das, Supratim ... 11
Das, Tapas K. ... 63
De Meyer, Frédérick ... 74
Dewan, Alim ... 52
Du, Chunxiao ... 117
Dunn, Jennifer B. ... 65, 113
Duoss, Eric B. ... 17
Fan, Senqing ... 69
Florian, Jacob ... 14
Floudas, Christodoulos A. ... 23
Freeman, Charles J. ... 65
Gao, Zhennan ... 124
Geng, Zhigang ... 124
Ghorbanian, Mahyar ... 67
Ghosh, Arpa ... 62
Goldsmith, Bryan R. ... 14, 59
Gribble, David ... 73
Guenther, Chris ... 18
Gunukula, Sampath ... 64
Guo, Lin ... 46
Guo, Ping ... 88
Guo, Xiaoyun ... 3
Hallermann, Detlef R. ... 23
Haugen, Stefan ... 61
He, Aaron ... 50
He, Qinggang ... 49
Hesselink, Lambertus ... 7
Huang, Baogang ... 124
Huang, Haoyong ... 115
Huang, Zhengliang ... 3
Hughes, Ryan ... 24
Hunt, Savannah G. ... 66
Hwang, Hyun-Tae ... 66
Hwang, Sungwon ... 44, 60
Itaya, Yoshinori ... 85
Jadhav, Ankur L. ... 10
Jay, Rahul ... 10
Jayaraman, Ambalavanan ... 73
Jin, Zhehui ... 87
Ju, Bin ... 88
Kalaga, Dinesh ... 134
Kamiya, Kenji ... 85
Kani, Nishithan C. ... 1
Kansha, Yasuki ... 83
Kariuki, Nancy N. ... 47
Karp, Eric M. ... 61
Kawaji, Masahiro ... 134
Kim, Byungmook ... 53
Kim, Geo-Jong ... 66
Ko, Eunbyeol ... 59
Kobayashi, Nobusuke ... 85
Kolhekar, Snehal ... 56
Koman, Volodymyr ... 58

Konz, Zachary 6, 55
Koronaios, Peter 110
Kotamreddy, Goutham....... 24
Kubic Jr., William L. 65
Kwon, Joseph....... 19, 44
Lamanna, Jacob 7
Lee, Dongcheul....... 53
Lee, Dongjun 60
Lee, Hyeonggeon....... 44
Lee, Jaewoo 53
Lemaire III, Peter....... 110
Li, Wenhui....... 87
Li, Xinhao....... 9
Liao, Zuwei....... 3
Liu, Albert Tianxiang 58
Liu, Jingyun....... 69
Liu, Jinlu....... 112
Liu, Weilin....... 124
Liu, Yong....... 18
Lowd, Jack....... 68
Lu, Shanfu 49
Lüönd, Lorenz 137
Manchenahalli, Manohar 67
Manousiouthakis, Ioannis 26
Manousiouthakis, Vasilios....... 26, 68
Matthews, Logan R....... 23
McCloskey, Bryan D. 6
McShane, Eric 6
Mei, Haiyan 88
Messinger, Robert....... 10, 54, 134, 135
Metz, Patrick....... 52
Mooney, Mason 47
Myers, Deborah J....... 47
Nakagawa, Tsuguhiko 85
Neyerlin, K. C....... 47
Nielsen, Caroline 43
Niziolek, Alexander M. 23
Noronha, Eyan 21
Nyce, Michael....... 54, 56
Onel, Onur 23
Parambathu, Arjun Valiya 112
Park, Jungeun....... 134, 135
Paul, Partha....... 7
Phirani, Jyoti 2
Pichardo, Patricia....... 26
Pistikopoulos, Efstratios N....... 23
Prajapati, Aditya 84
Radke, Clayton J....... 50
Rashid, Suliman....... 72
Rosenthal, Alex....... 65
Saboe, Patrick 61
Sato, Yuki 83
Shahnam, Mehrdad 18
Shin, Chee Burm....... 53
Shuai, Yun....... 3
Singh, Deepshikha 2
Singh, Meenesh R. 1, 84
Singh, Nirala 14, 59
Sishtla, Chakravarthy 73
Sitapure, Niranjan 44
Smith, Joseph D. 67
Smith, Kandler 6
Smith, Ryan G....... 62
Son, Sang Hwan....... 19
Song, Yi 115
Sorescu, Sorin M....... 23
Stanford, John P. 62
Steinfeld, Aldo 137
Stracensky, Thomas 46
Strano, Michael 58
Su, Yanchun....... 124
Sun, Jingyuan....... 3
Sweeny, William....... 52
Tan, Eric....... 65
Tezel, F. Handan 136
Toney, Michael 7
Tongwen, Jiang 116
Tortorelli, Daniel A....... 17
Tran, Dat T....... 64
Tsotsis, Theodore T....... 68
Turney, Damon E....... 54
Urban, Alexander....... 9
Valentino, Lauren....... 65
Van Cleve, Timothy 47
Wang, Guanxiong 47
Wang, Jingdai....... 3
Wang, Yun 46
Watts, Seth E....... 17
Weber, Adam Z....... 50
Wei, Shuya....... 5
Wei, Xie 116
Weker, Johanna....... 7
Wild, Michael....... 137
Wong, Jonathan M. 17
Worsley, Marcus 17
Wu, Gang 46, 48
Wu, Jianfa 115
Xiao, Zeyi 69
Xu, Hui....... 46
Xueming, Dong....... 61
Yadav, Gautam G....... 54
Yang, Fan....... 46
Yang, Jun 117, 124
Yang, Yao 3

Yang, Yongrong 3
Ydstie, B. Erik 21
Yong, Wang..116
Yu, Jia..67
Yusuf, Maha 7
Zeng, Fanhua 88, 117, 124
Zhang, Ge 58
Zhang, Jujia 49
Zhang, Maolin 88
Zhang, Shuomeng..49
Zhang, Yizhong 88
Zhao, Shengxian 115
Zhong, Sichen..46
Zych, Artur 134